McGraw-Hill Education

500 Review Questions for the MCAT: Organic Chemistry and Biochemistry

Also in McGraw-Hill's 500 Questions Series

McGraw-Hill's 500 American Government Questions: Ace Your College Exams
McGraw-Hill's 500 College Algebra and Trigonometry Questions: Ace Your College Exams
McGraw-Hill's 500 College Biology Questions: Ace Your College Exams
McGraw-Hill's 500 College Calculus Questions: Ace Your College Exams
McGraw-Hill's 500 College Chemistry Questions: Ace Your College Exams
McGraw-Hill's 500 College Physics Questions: Ace Your College Exams
McGraw-Hill's 500 Differential Equations Questions: Ace Your College Exams
McGraw-Hill's 500 European History Questions: Ace Your College Exams
McGraw-Hill's 500 French Questions: Ace Your College Exams
McGraw-Hill's 500 Linear Algebra Questions: Ace Your College Exams
McGraw-Hill's 500 Macroeconomics Questions: Ace Your College Exams
McGraw-Hill's 500 Microeconomics Questions: Ace Your College Exams
McGraw-Hill's 500 Organic Chemistry Questions: Ace Your College Exams
McGraw-Hill's 500 Philosophy Questions: Ace Your College Exams
McGraw-Hill's 500 Physical Chemistry Questions: Ace Your College Exams
McGraw-Hill's 500 Precalculus Questions: Ace Your College Exams
McGraw-Hill's 500 Psychology Questions: Ace Your College Exams
McGraw-Hill's 500 Spanish Questions: Ace Your College Exams
McGraw-Hill's 500 U.S. History Questions, Volume 1: Ace Your College Exams
McGraw-Hill's 500 U.S. History Questions, Volume 2: Ace Your College Exams
McGraw-Hill's 500 World History Questions, Volume 1: Ace Your College Exams
McGraw-Hill's 500 World History Questions, Volume 2: Ace Your College Exams

McGraw-Hill's 500 MCAT Biology Questions to Know by Test Day
McGraw-Hill's 500 MCAT General Chemistry Questions to Know by Test Day
McGraw-Hill's 500 MCAT Physics Questions to Know by Test Day

McGraw-Hill Education
500 Review Questions for the MCAT: Organic Chemistry and Biochemistry

John T. Moore, EdD
Richard H. Langley, PhD

New York Chicago San Francisco Athens London Madrid
Mexico City Milan New Delhi Singapore Sydney Toronto

Copyright © 2015 by The McGraw-Hill Education, Inc. All rights reserved. Printed in the United States of America. Except as permitted under the United States Copyright Act of 1976, no part of this publication may be reproduced or distributed in any form or by any means, or stored in a database or retrieval system, without the prior written permission of the publisher.

1 2 3 4 5 6 7 8 9 10   QFR/QFR   1 2 1 0 9 8 7 6 5

ISBN   978-0-07-183486-5
MHID   0-07-183486-9

e-ISBN   978-0-07-183481-0
e-MHID 0-07-183481-8

Illustrations by Cenveo

MCAT is a registered trademark of the Association of American Medical Colleges, which was not involved in the production of, and does not endorse, this product.

McGraw-Hill Education products are available at special quantity discounts to use as premiums and sales promotions or for use in corporate training programs. To contact a representative, please e-mail us at bulksales@mcgraw-hill.com.

This book is printed on acid-free paper.

CONTENTS

Introduction vii

Chapter 1 **The Fundamentals** 1
Questions 1–53

Chapter 2 **Isomers and Physical Properties** 23
Questions 54–116

Chapter 3 **Substitution and Elimination Reactions** 49
Questions 117–179

Chapter 4 **Electrophilic Addition Reactions** 75
Questions 180–242

Chapter 5 **Nucleophilic and Cyclo Addition Reactions** 109
Questions 243–304

Chapter 6 **Lab Technique and Spectroscopy** 143
Questions 305–366

Chapter 7 **Bioorganic Chemistry** 163
Questions 367–429

Chapter 8 **Final Review** 195
Questions 430–500

Answers 223

ABOUT THE AUTHORS

John T. Moore, EdD, has taught chemistry at Stephen F. Austin State University for more than 40 years, where he is director of the Teaching Excellence Center and codirector of the Science, Technology, Engineering and Mathematics Center.

Richard H. Langley, PhD, has taught chemistry at the university level for more than 30 years and has coauthored numerous books on the subject. He has written questions for the AP Chemistry exam and taught general chemistry and organic chemistry MCAT review courses.

INTRODUCTION

Congratulations! You've taken a big step toward MCAT success by purchasing *500 Review Questions for the MCAT: Organic Chemistry and Biochemistry*. We are here to help you take the next step and score high on the MCAT so that you can get into the medical school of your choice.

This book gives you 500 multiple-choice questions that cover all the most essential course material. Each question is clearly explained in the answer key. The questions will give you valuable independent practice to supplement your regular textbook and the ground you have already covered in your classes.

This book and the others in the series were written by expert teachers who know the MCAT inside and out and can identify crucial information as well as the kinds of questions that are most likely to appear on the exam.

You might be the kind of student who needs to do some extra studying a few weeks before the exam for a final review. Or you might be the kind of student who puts off preparing until the last minute before the exam. No matter what your preparation style, you will benefit from reviewing these 500 questions, which parallel the content and degree of difficulty of the questions on the actual MCAT. These questions and the explanations in the answer key are the ideal last-minute study tool for those final weeks before the test.

If you practice with all the questions and answers in this book, we are certain you will build the skills and confidence you need to excel on the MCAT. Good luck!

—*The Editors of McGraw-Hill Education*

CHAPTER 1

The Fundamentals

Questions 1–7 refer to the following passage.

Use of the valence bond theory begins with the s, p, and d atomic orbitals and redistributes electrons to give a more favorable (lower-energy) arrangement. The rearrangement of the electron configuration leads to the formation of new orbitals, which can then overlap and share electrons. *Valence bond theory* uses hybridization to explain molecular geometry. *Hybridization* results from the combination of atomic orbitals into new hybrid orbitals. When atomic orbitals (s, p, and d) hybridize, both the shape and the label of the orbitals change. Hybridization never changes the number of orbitals, however, so that the hybridization of four atomic orbitals must yield four hybrid orbitals.

The hybrid orbitals that form, along with the atomic orbitals that remain unhybridized, tell us many different things about the molecules in addition to the molecular geometry. For example, suppose the geometry of BeI_2 is to be determined. The electron configuration of the beryllium atom is $1s^2 2s^2$ or $[He]2s^2$. One of these valence 2s electrons will be shared with one of the iodine atoms, and the second one will be shared with the other iodine atom. If we pause and reexamine the initial Lewis structure of beryllium iodide, we find that the beryllium electrons are separated, not paired, as they must be in an s orbital. Thus, to achieve the beryllium electron arrangement indicated by the Lewis structure, it is necessary to rearrange the electrons in order to account for this fact. The valence shell of beryllium contains 2p orbitals in addition to the 2s orbitals. It is possible to move one of the 2s electrons of the beryllium atom into one of the 2p orbitals. The resultant excited-state beryllium atom is less stable than the original ground-state beryllium atom.

The occupied orbitals will hybridize (blend together), and the empty orbitals will remain as normal 2p orbitals. The two orbitals (the 2s orbital and the one occupied 2p orbital) will produce two hybrid orbitals. These hybrid orbitals are both labeled sp hybrids because they form from an s orbital and a p orbital. The separated beryllium electrons are now free to pair with one iodine electron each to form two covalent bonds. Hybrid orbitals contain features of each "parent" orbital. Each of the sp hybrids has half s character and half p character.

1. The following illustration is a section of a carbon nanotube.

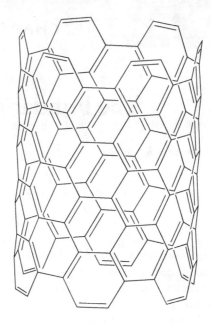

What is the best description of the hybridization of the carbon atoms?
(A) sp³ only
(B) sp² only
(C) sp² and sp³
(D) sp and sp²

2. What is the hybridization of the nitrogen atoms in the following compound?

(A) Both are sp³ hybridized.
(B) Both are sp hybridized.
(C) One is sp hybridized, and the other is sp³ hybridized.
(D) One is sp hybridized, and the other is sp² hybridized.

3. What are the hybridizations of the indicated atoms?

(A) 1 = sp^2, 2 = sp^2, and 3 = sp^2
(B) 1 = sp, 2 = sp, and 3 = sp^2
(C) 1 = sp^3, 2 = sp^3, and 3 = sp^2
(D) 1 = sp, 2 = sp^3, and 3 = sp^2

4. The bonds between the carbon atoms in 1-butyne utilize which of the following types of orbitals?
(A) sp^2 and sp^3 hybridized orbitals
(B) sp and sp^2 hybridized orbitals
(C) sp and sp^3 hybridized orbitals
(D) sp hybridized orbitals

5. The structure of the drug Demerol is:

How many sp³ hybridized carbon atoms are present in Demerol?
(A) 4
(B) 5
(C) 6
(D) 8

6. What is the hybridization on each of the indicated nitrogen atoms?

(A) I is sp, II is sp³, and III is sp².
(B) I is sp, II is sp², and III is sp.
(C) I is p, II is sp², and III is sp².
(D) I is sp³, II is sp², and III is sp².

7. Which orbitals interact when CH_3^+ reacts with CH_3^- to form CH_3CH_3?
   (A) An unhybridized p orbital on the CH_3^+ and an sp³ hybridized orbital on the CH_3^-
   (B) An sp² hybridized orbital on the CH_3^+ and an sp² hybridized orbital on the CH_3^-
   (C) An sp³ hybridized orbital on the CH_3^+ and an sp³ hybridized orbital on the CH_3^-
   (D) An unhybridized s orbital on the CH_3^+ and an unhybridized p orbital on the CH_3^-

8. Which of the following is the best resonance structure for HOCN?

(A) 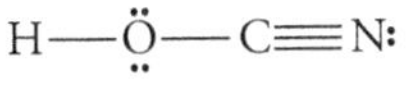 H—Ö—C≡N:

(B) H—O⁺=C=N:⁻

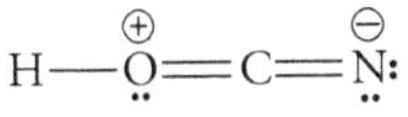

(C) H—Ö=C=N:

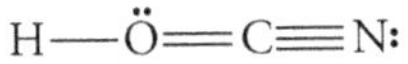

(D) 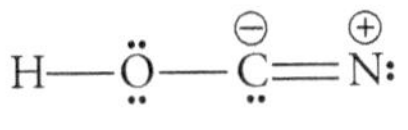 H—Ö—C⁻=N:⁺

9. Which of the following are reasonable resonance structures for nitrobenzene?

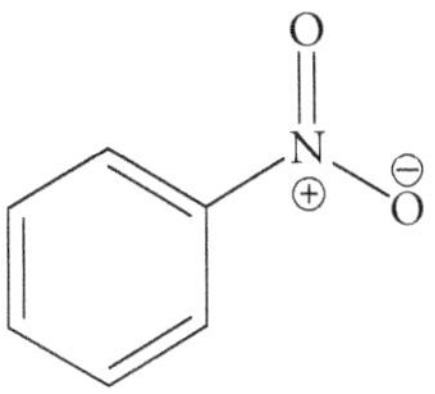

I 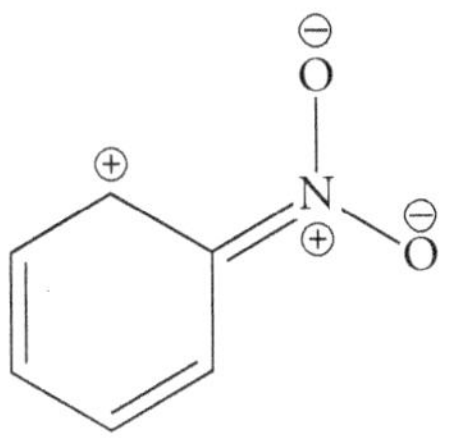   II 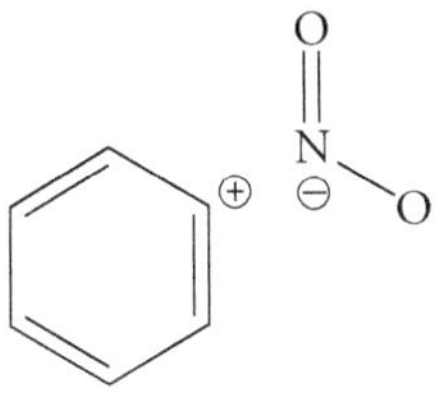   III

(A) I, II, and III
(B) I and III
(C) II and III
(D) I and II

10. Choose the strongest acid from the following:

(A) 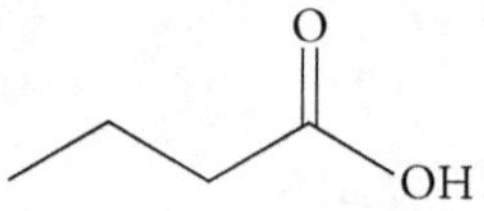 CH₃CH₂CH₂COOH

(B) 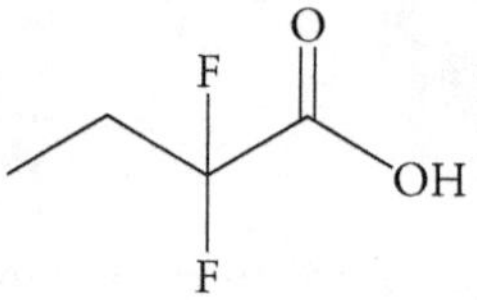 CH₃CH₂CF₂COOH

(C) 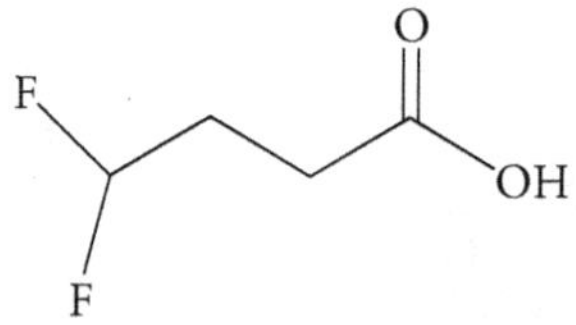 CHF₂CH₂CH₂COOH (with F's on terminal carbon)

(D) 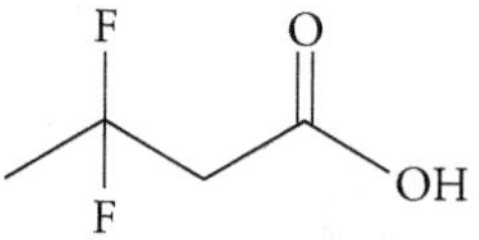 CF₂(CH₃)CH₂COOH

11. Why is ortho-nitrophenol a stronger acid than phenol?
    (A) Destabilization of the aromatic system
    (B) Electron donation by the nitro group
    (C) Steric crowding between the –OH and the –NO₂
    (D) Resonance stabilization of the anion

12. Which of the following is NOT a disubstituted cyclohexane?

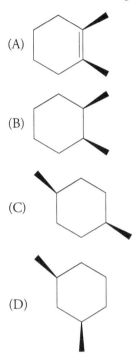

13. Why is para-aminophenol a weaker acid than phenol?
    (A) The amino group is larger than hydrogen, and therefore it creates more steric hindrance.
    (B) The amino group is an electron-donating group, which stabilizes the conjugate base.
    (C) The amino group destabilizes the aromatic system.
    (D) The amino group is an electron-donating group, which destabilizes the conjugate base.

38. Two cysteine molecules in a protein may undergo oxidation to form a disulfide linkage. It is possible to reverse this process by adding a suitable reducing agent. The process is:

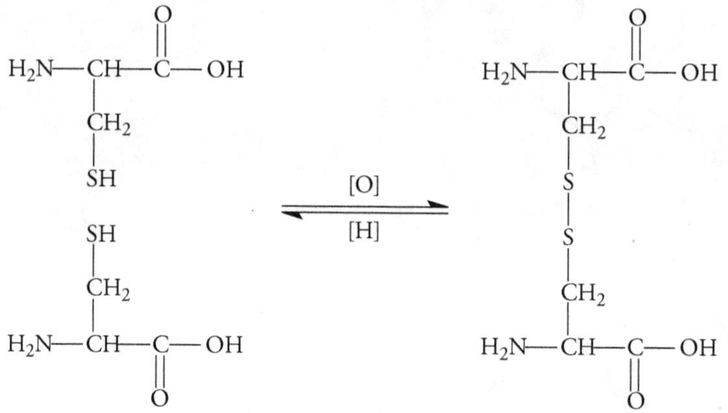

What is the change in the oxidation state of the sulfur atoms during reduction?

(A) −1 to −2
(B) 2 to −1
(C) 2 to 0
(D) 1 to 0

39. What is the name of the following compound?

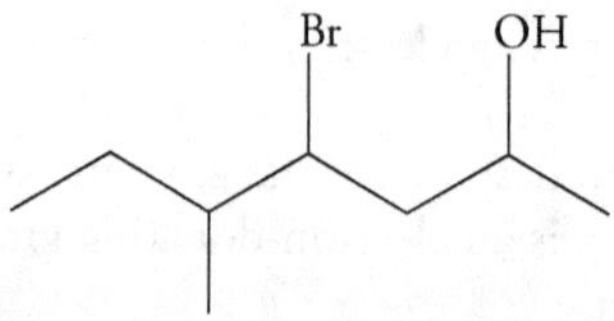

(A) 4-bromo-6-hydroxy-3-methylheptane
(B) 4-bromo-3-methyl-heptan-6-ol
(C) 4-bromo-2-hydroxy-5-methylheptane
(D) 4-bromo-5-methyl-heptan-2-ol

18. In the following ion, rank the indicated bonds from longest to shortest.

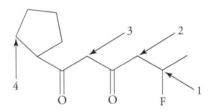

(A) 3 > 1 = 2
(B) 1 > 3 = 2
(C) 3 = 2 > 1
(D) 1 = 2 > 3

19. Which of the indicated hydrogen atoms in the following compound is the most acidic?

(A) 3
(B) 1
(C) 2
(D) 4

20. The following molecule contains both σ and π bonds. How many σ bonds are present?

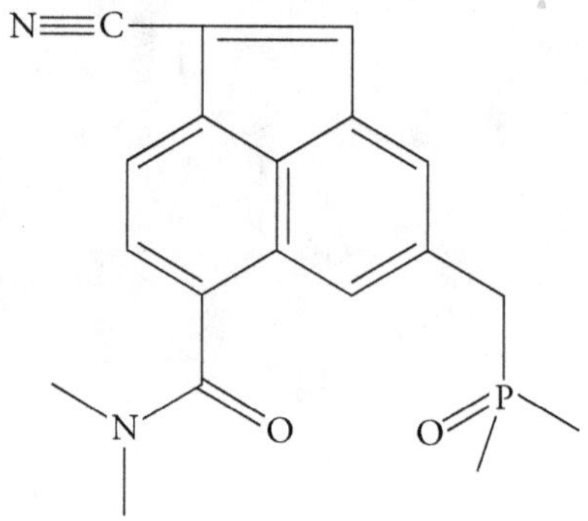

(A) 23
(B) 32
(C) 22
(D) 42

21. The following molecule does NOT contain which of the following functional groups?

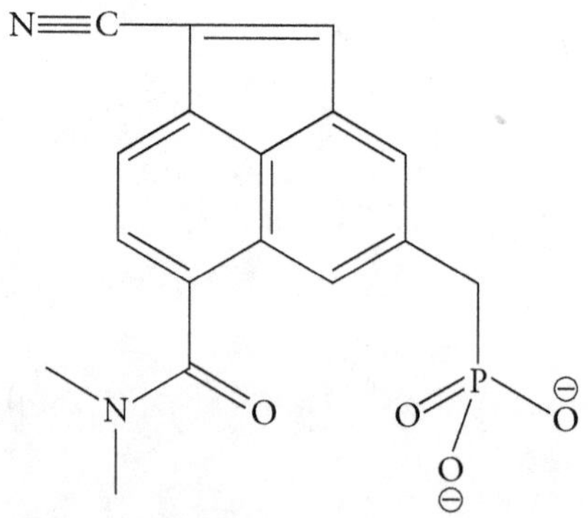

(A) Amide
(B) Nitrile
(C) Amine
(D) Phosphate

22. What is the IUPAC name for the following compound?

22. What is the IUPAC name for the following compound?

(A) 5-hexyn-3-ol
(B) 1-hexyn-4-ol
(C) cis-5-hexyn-3-ol
(D) 4-hydroxy-1-hexyne

23. Is the following compound acidic, basic, or neutral in aqueous solution?

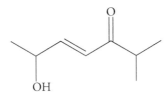

(A) Neutral
(B) Basic
(C) Amphoteric
(D) Acidic

24. How many σ bonds and π bonds are in naphthalene?
(A) 11 σ and 5 π
(B) 19 σ and 5 π
(C) 21 σ and 5 π
(D) 8 σ and 4 π

25. During the complete hydrogenation of one mole of $C_{10}H_{14}$, three moles of hydrogen were absorbed. How many rings were present in the original compound?
(A) 1
(B) 2
(C) 3
(D) 4

26. A compound with the general formula $C_{10}H_{23}NO$ could NOT contain which of the following functional groups?
    (A) Ether
    (B) Alcohol
    (C) Amide
    (D) Amine

27. Which of the following series best relates the order of decreasing C-O bond length?
    (A) $CO > CO_2 > CO_3^{2-} > CH_3OH$
    (B) $CH_3OH > CO_3^{2-} = CO_2 > CO$
    (C) $CH_3OH > CO_3^{2-} > CO_2 = CO$
    (D) $CH_3OH > CO_3^{2-} > CO_2 > CO$

28. Looking down the C_1-C_2 bond, how do the stabilities of the different conformers of 1-chloropropane compare?
    (A) anti < eclipsed < gauche
    (B) eclipsed < gauche < anti
    (C) gauche < eclipsed < anti
    (D) anti < gauche < eclipsed

29. What is the IUPAC name for the following compound?

    (A) (Z)-4-heptene
    (B) (Z)-3-heptene
    (C) (E)-3-heptene
    (D) (E)-4-heptene

30. How many stereoisomers does the following compound have?

    (A) 16
    (B) 4
    (C) 2
    (D) 8

31. What is the name of the following compound?

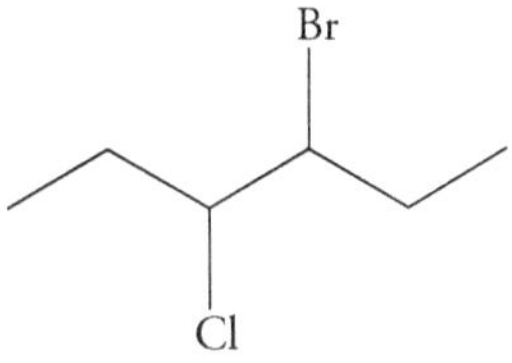

(A) 3-bromo-4-chlorohexane
(B) trans-3-bromo-4-chlorohexane
(C) E-3-bromo-4-chlorohexane
(D) 1-bromo-2-chloro-1,2-diethylethane

32. Which of the following is an acid anhydride?

(A)

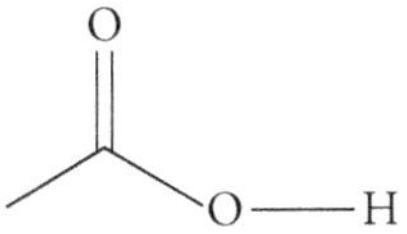

(B)

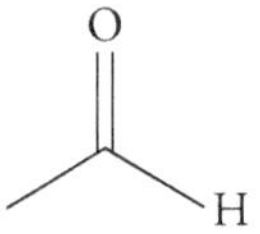

(C)

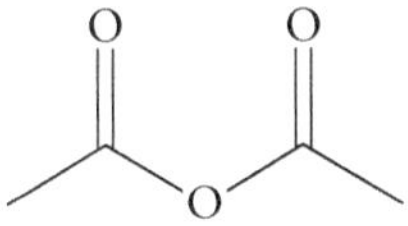

(D)

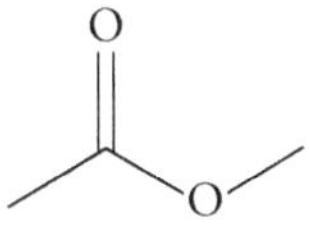

33. How many isomers of hexane are there?
(A) 7
(B) 6
(C) 5
(D) 4

34. Which of the following is an isobutyl group?

(A) 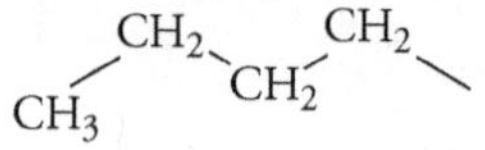 CH₃–CH₂–CH₂–CH₂–

(B) 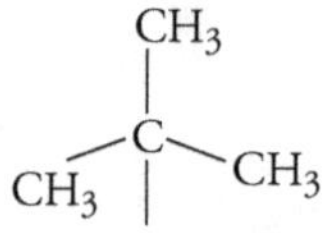 (CH₃)₃C–

(C) 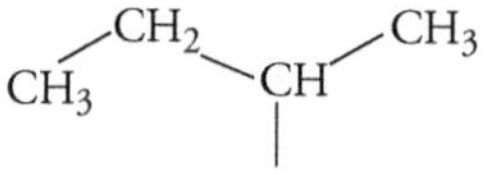 CH₃–CH₂–CH(CH₃)–

(D) 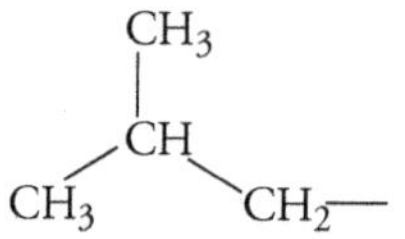 (CH₃)₂CH–CH₂–

35. What is the IUPAC name for the drug Tylenol, shown as follows?

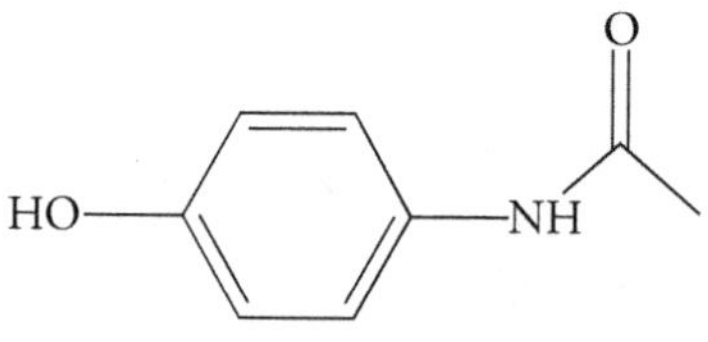

(A) N-(4-hydroxyphenyl)acetamide
(B) 4-(N-aceto)phenol
(C) N-(4-phenol)-N-acetoamine
(D) N-(p-phenol)acetamide

36. What is the IUPAC name for the following compound?

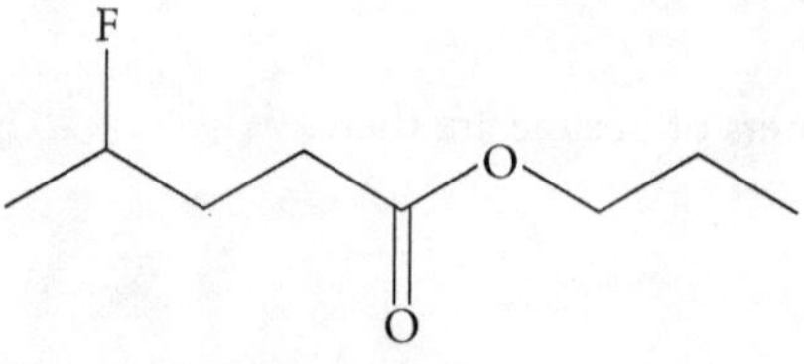

(A) 2-fluoro-5-nonanone
(B) propyl 4-fluoropentanoate
(C) 4-fluoropentyl propyl ether
(D) 4-fluoropentyl propanoate

37. Which of the following compounds has four degrees of unsaturation?

(A) 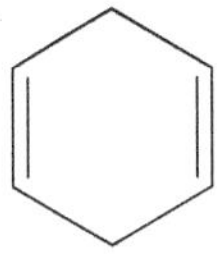 [cyclohexene]

(B) 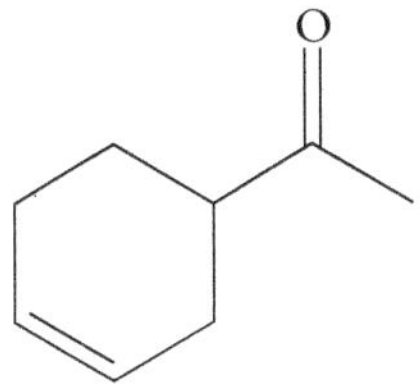 [cyclohexene with COCH3 substituent]

(C) 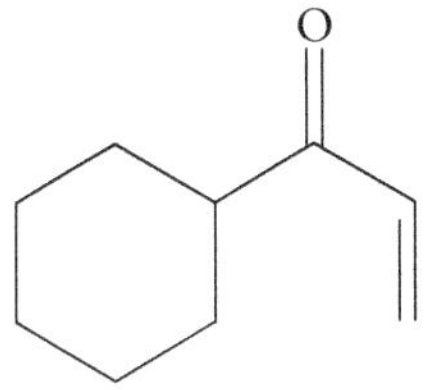 [cyclohexane with COCH=CH2 substituent]

(D) 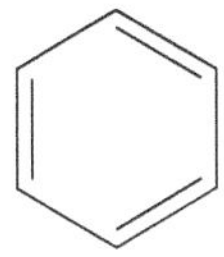 [benzene]

38. Two cysteine molecules in a protein may undergo oxidation to form a disulfide linkage. It is possible to reverse this process by adding a suitable reducing agent. The process is:

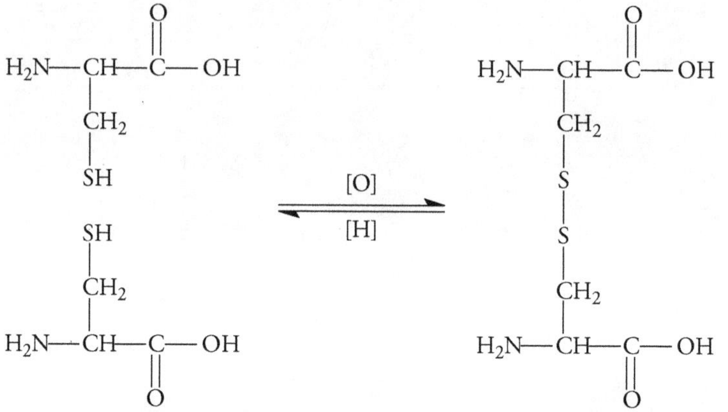

What is the change in the oxidation state of the sulfur atoms during reduction?

(A) 1 to –2
(B) 2 to –1
(C) 2 to 0
(D) 1 to 0

39. What is the name of the following compound?

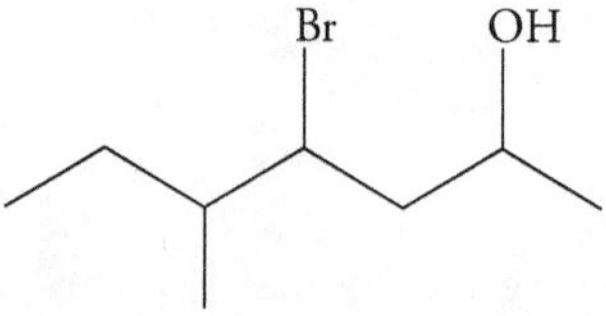

(A) 4-bromo-6-hydroxy-3-methylheptane
(B) 4-bromo-3-methyl-heptan-6-ol
(C) 4-bromo-2-hydroxy-5-methylheptane
(D) 4-bromo-5-methyl-heptan-2-ol

40. Which of the following is an acetal?

(A) 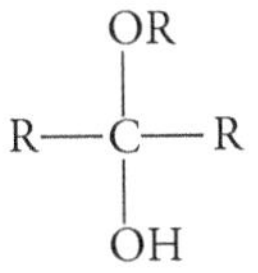 R—C(OR)(OH)—R

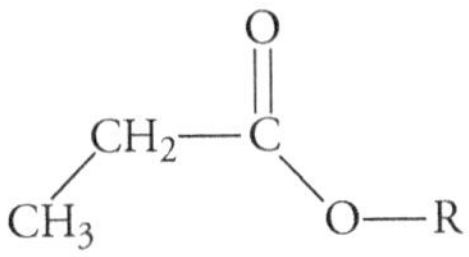

(B) CH₃—CH₂—C(=O)—O—R

(C) R—C(=O)—NH₂

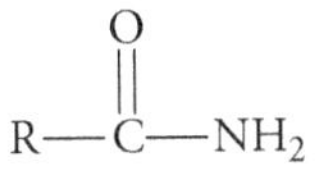

(D) 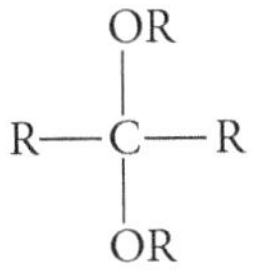 R—C(OR)(OR)—R

41. What is the IUPAC name for the following compound?

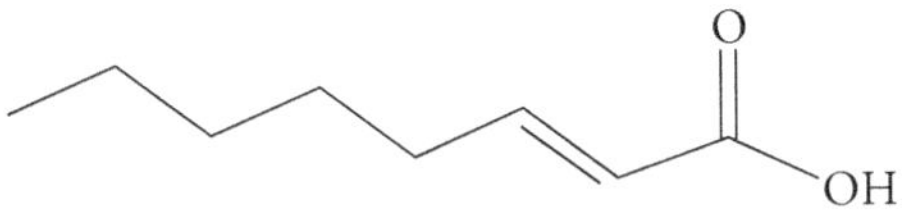

(A) (E)-6-octenoic acid
(B) (Z)-2-octenoic acid
(C) (E)-2-octenoic acid
(D) (Z)-6-octenoic acid

42. How many different isomers of $C_3H_4F_2$ are there?
(A) 10
(B) 7
(C) 3
(D) 8

43. Distillation of crude oil yields many compounds. One of the compounds is $C_{12}H_{14}$. What is the degree of unsaturation of this compound?
(A) 10
(B) 8
(C) 4
(D) 6

44. The compound C_3H_5Br could be any of the following EXCEPT:
    (A) trans-2-bromopropene
    (B) bromocyclopropane
    (C) trans-1-bromopropene
    (D) cis-1-bromopropene

45. Which of the isomers of isopropylcyclohexane is the most stable?
    (A) Axial
    (B) Equatorial
    (C) Cis
    (D) Trans

46. The free radical monochlorination of 2-methylbutane will give the following products.

    What is the relationship between these products?
    (A) Stereoisomers
    (B) Conformers
    (C) Structural isomers
    (D) Diastereomers

47. Which of the following is the most stable hydrocarbon?

(A)

(B)

(C)

(D)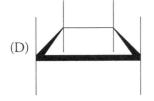

48. Rank the following acids from strongest to weakest:

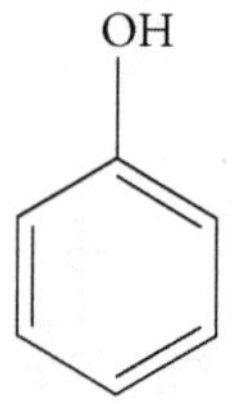

I: phenol (C₆H₅OH)

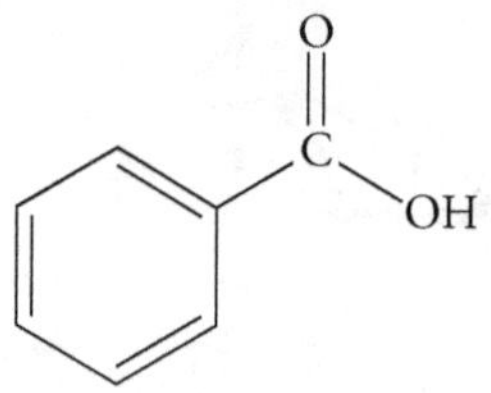

II: benzoic acid (C₆H₅COOH)

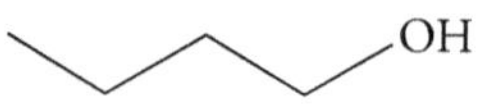

III: butanol (CH₃CH₂CH₂CH₂OH)

IV: H-Cl

(A) IV > II > I > III
(B) III > I > II > IV
(C) IV > II > III > I
(D) II > I > III > IV

49. What are the approximate values of each of the indicated bond angles in the structure of the amino acid histidine?

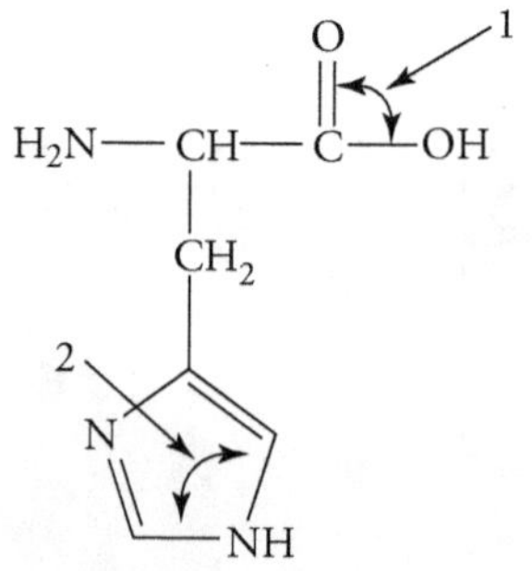

H₂N—CH—C(=O)—OH, with CH₂ substituent connected to imidazole ring (N, NH); angle 1 indicated at C(=O)–OH; angle 2 indicated at ring N.

(A) 1 = 109.5° and 2 = 109.5°
(B) 1 = 90° and 2 = 120°
(C) 1 = 120° and 2 = 109.5°
(D) 1 = 109.5° and 2 = 120°

50. The following compounds are examples of:

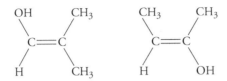

(A) Enantiomers.
(B) Conformers.
(C) Regioisomers.
(D) Diastereomers.

51. The compound 1,3,5-hexatriene contains a conjugated system. All the C-C single bonds are shorter than normal C-C single bonds. The reason for this shortening is that there is:

(A) Double-bond character due to resonating σ-electrons.
(B) Partial overlap of the unhybridized p orbitals on the two carbon atoms.
(C) An interaction between the sp² and sp³ hybridized orbitals on the two carbon atoms.
(D) Bond shortening when two sp³ hybridized orbitals overlap.

52. The compound 3-hexene exists as cis and trans isomers. Which has the greater heat of hydrogenation?

(A) Both isomers have the same heat of hydrogenation.
(B) The trans isomer has the greater heat of hydrogenation.
(C) The cis isomer has the greater heat of hydrogenation.
(D) It is impossible to tell from the data in the problem.

53. Determine the degree of unsaturation in $C_6H_{10}N_2O$.

(A) 3
(B) 2
(C) 1
(D) 0

CHAPTER 2

Isomers and Physical Properties

Questions 54–60 refer to the following passage.

Isomers are compounds that have the same molecular formula, but different structural formulas. Many organic and biochemical compounds may exist in different isomeric forms. Many times, especially in biological systems, these different isomers have different properties. The presence of an asymmetric, or *chiral*, carbon atom is sufficient to produce isomeric molecules. A chiral carbon atom has four different groups attached to it. The majority of biological molecules have one or more chiral carbon atoms, and, for this reason, they are chiral.

These optical isomers are important in living cells, where one form may be biologically active and the other inactive. Many drugs exist in optically active forms. One form may have the desired results, while the other may produce an entirely different set of biological results. The classic case is the drug thalidomide, in which one isomer acts as an antidepressant, while the other causes mutations in fetuses.

54. Why is the following compound achiral (does not exhibit optical activity)?

(A) The Me and Et groups are too similar.
(B) The compound is meso.
(C) Nitrogen compounds are always achiral.
(D) The nitrogen undergoes inversion.

55. The following compound has two chiral carbons. What are their absolute configurations?

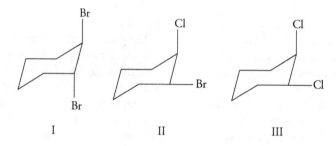

(A) 2R, 3R
(B) 2R, 3S
(C) 2S, 3S
(D) 2S, 3R

56. Which of the following compounds is achiral?

(A) III
(B) II
(C) I
(D) I and III

57. The following molecule is chiral, as is the corresponding fluorine derivative. How does the optical rotation of this compound compare to that of the corresponding fluorine derivative?

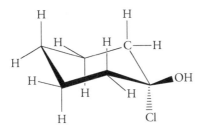

(A) One is (+) and one is (–).
(B) Both are (+).
(C) Both are (–).
(D) It is impossible to determine.

58. The amino acid glycine, unlike most natural amino acids, is achiral. The laboratory synthesis of glycine produces pure glycine; however, the laboratory syntheses of the other natural amino acids are only 50% pure. What is the name of the 50% pure product?
(A) An optically active mixture
(B) A diastereomeric mixture
(C) A mesomeric mixture
(D) A racemic mixture

59. The compound 2,3-dihyroxypentane has how many chiral centers?
(A) 1
(B) 2
(C) 0
(D) 3

60. A chiral compound with an optical rotation of +15.6° undergoes an S_N2 reaction at the chiral center to produce another chiral compound. What would be the optical rotation of the product?
(A) It would not be possible to predict the optical rotation.
(B) The optical rotation would remain at +15.6°.
(C) The optical rotation would change to –15.6°.
(D) The optical rotation would become 0°.

61. Resonance stabilizes many compounds. The following is an example of resonance.

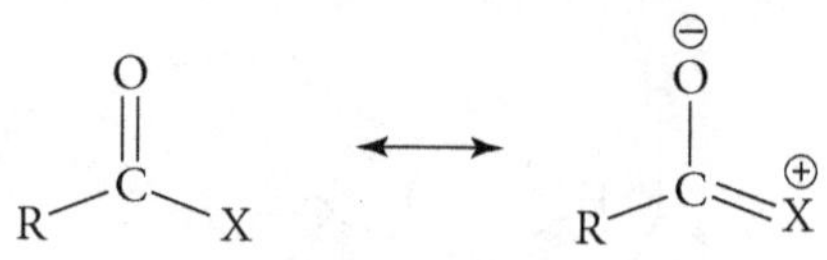

$$R-\underset{X}{\overset{O}{\underset{\|}{C}}} \longleftrightarrow R-\underset{X^{\oplus}}{\overset{O^{\ominus}}{C}}$$

Which of these compounds would benefit the most from this type of resonance?

(A) Amide
(B) Ester
(C) Carboxylic acid
(D) Acid chloride

62. Which of the following is (1R,3S)-1,3-dibromo-1,3-difluorobutane?

(A) [structure with Br and F on left C (F up, Br back, H down), CH$_2$ bridge, right C with F forward, Br up, CH$_3$]

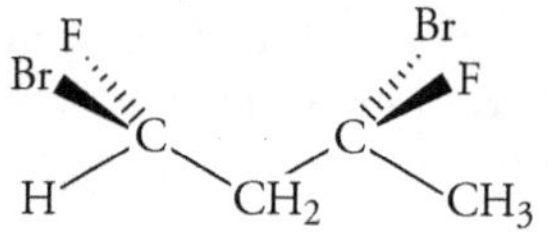

(B) [structure with F and Br on left C (Br up, F back, H down), CH$_2$ bridge, right C with F up, Br forward, CH$_3$]

(C) [structure with F and Br on left C (Br up, F back, H down), CH$_2$ bridge, right C with F up, Br forward, CH$_3$]

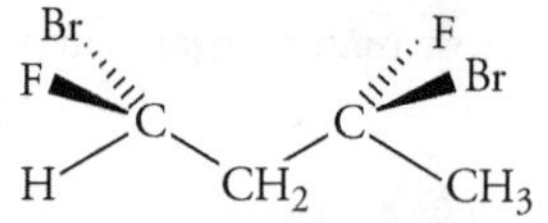

(D) [structure with Br and F on left C (F up, Br back, H down), CH$_2$ bridge, right C with F up, Br forward, CH$_3$]

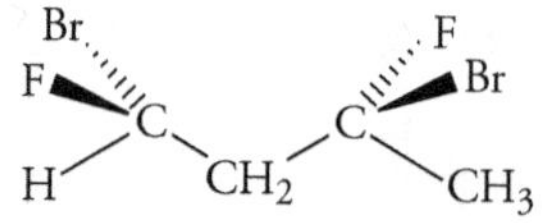

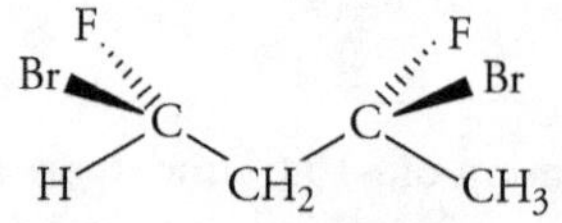

63. What is the relationship between the following two compounds?

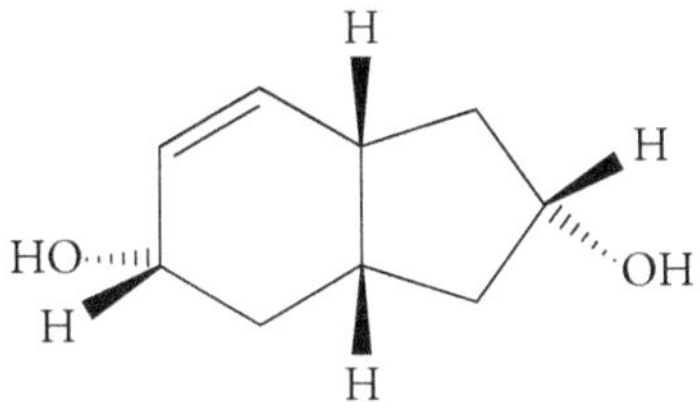

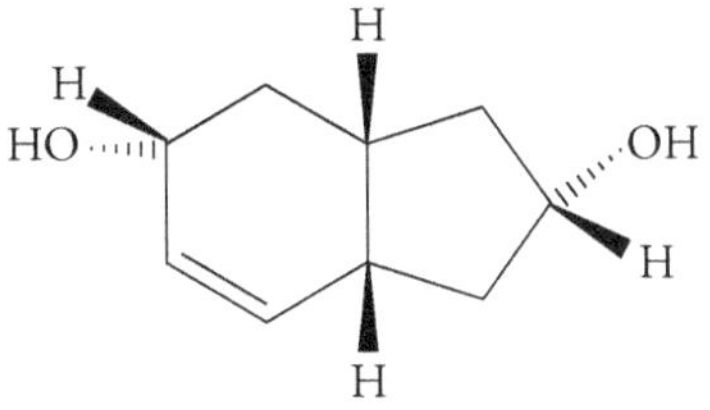

(A) Identical
(B) Enantiomers
(C) Diastereomers
(D) Conformers

64. Which of the following compounds is NOT polar?

(A)

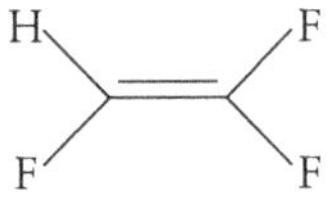

(B)

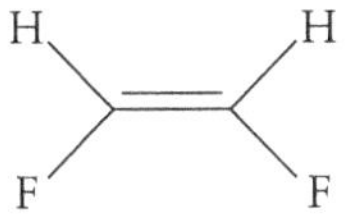

(C)

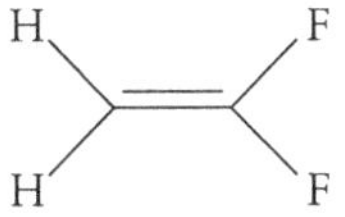

(D)

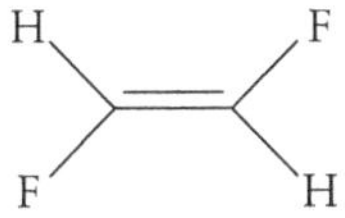

65. Why is the melting point of ethyl methyl amine higher than that of trimethyl amine?
   (A) Ethyl methyl amine has intermolecular hydrogen bonding.
   (B) Ethyl methyl amine does not have intermolecular hydrogen bonding.
   (C) Ethyl methyl amine has a higher molar mass.
   (D) Trimethyl amine has a higher molar mass.

66. What is the stereochemistry of the following cyclopropane?

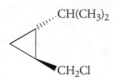

(A) 1S, 2R
(B) 2R, 3S
(C) 1S, 2S
(D) 2S, 3S

67. The last step in the synthesis of a new experimental drug involves the reaction of two optically active species. Even though each species contains only one chiral center, both species are available only as part of a racemic mixture. What is the maximum number of stereoisomers that the chemists will need to separate?

(A) 6
(B) 2
(C) 4
(D) 8

68. The S stereoisomer of a compound reacts with a racemic mixture of another compound. This results in two compounds that are easily separated from the reaction mixture. These two compounds are:
(A) Tautomers.
(B) Enantiomers.
(C) Conformers.
(D) Diastereomers.

69. Which of the following compounds have a nonzero net dipole moment?

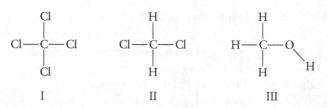

(A) I only
(B) III only
(C) II and III
(D) I and III

70. Arrange the following compounds in order of increasing dipole moment.

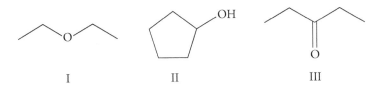

(A) I < II < III
(B) III < II < I
(C) I < III < II
(D) III < I < II

71. Which of the following compounds is expected to be the most soluble in water?

(A) HO~~~OH

(B) ~~~

(C) H₂N~~~NH₂

(D) Cl~~~Cl

72. Which of the following is capable of hydrogen bonding to water?

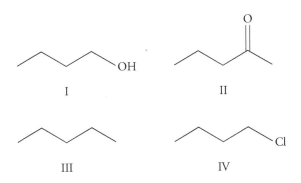

(A) IV only
(B) III and IV
(C) I only
(D) I and II

73. Which of the following has an S stereocenter?

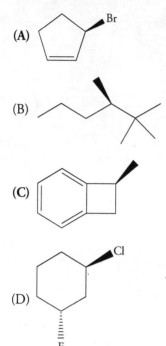

74. Which of the following is an enantiomer of this compound?

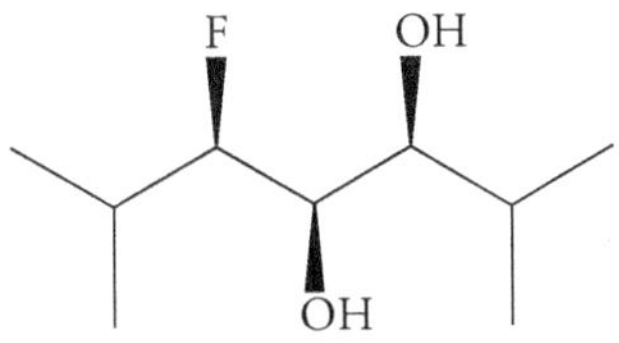

(A)

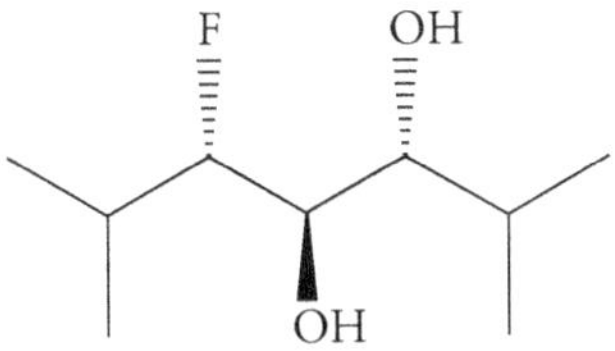

(B)

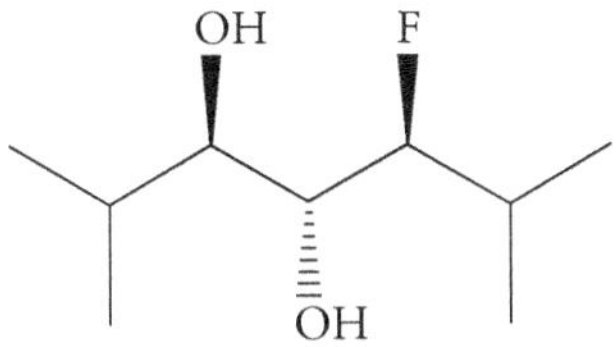

(C)

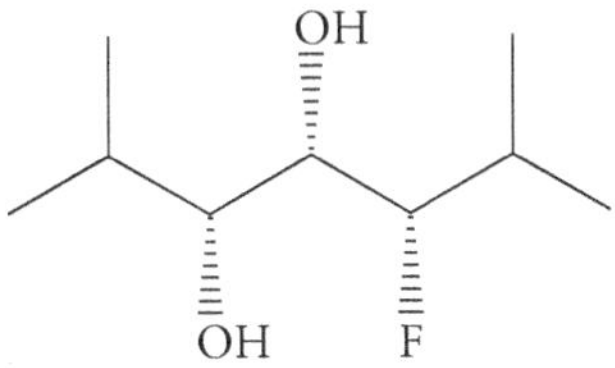

(D)

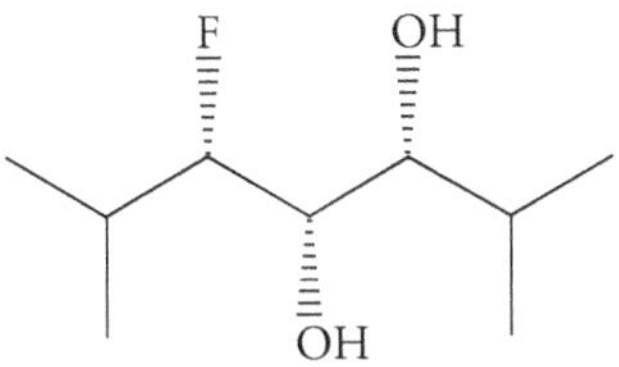

75. Which of the following is NOT a meso compound?

(A) [structure: cyclopentane with two OH groups, cis]

(B) [structure: cyclohexane with OH, two Cl, and CH₃ substituents]

(C) [structure: cyclohexane with three Cl substituents]

(D) [structure: BrCHF-CHFBr with stereochemistry shown]

76. Which of the following has an (R) stereocenter?

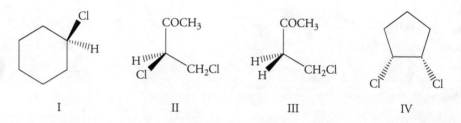

I   II   III   IV

(A) I
(B) III
(C) II
(D) IV

77. How do (+) and (−) symbols relate to d and l symbols?
    (A) (+) = l and (−) = d.
    (B) (+) = d and (−) = l.
    (C) It is necessary to check the structure of each form.
    (D) There is no relation.

78. Which of the following compounds is expected to be the most soluble in water?

    (A) 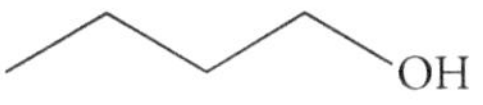 ~~~~OH

    (B) Na_2SO_4

    (C)  H_2N~~~~NH_2

    (D) 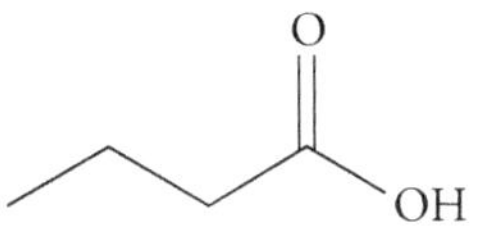 ~~~C(=O)OH

79. What is the best explanation of why compound I has a higher boiling point than compound II?

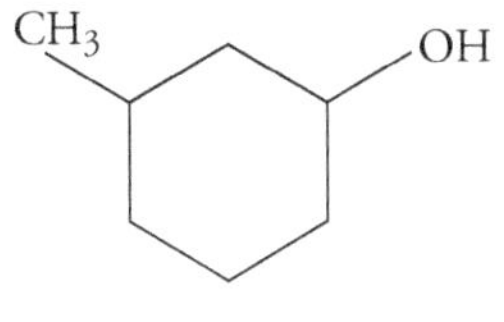

    I: 3-methylcyclohexanol   II: 1,4-dioxane

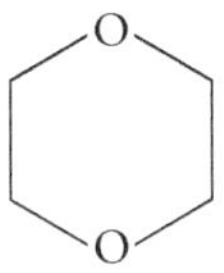

    (A) Compound II is unstable.
    (B) Compound II is nonpolar.
    (C) Compound I can ionize.
    (D) Compound I has hydrogen bonding.

80. What is the relationship between the following two molecules?

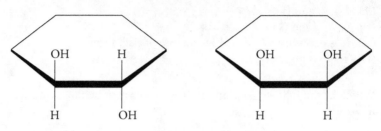

(A) Diastereomers
(B) Enantiomers
(C) Conformers
(D) Identical

81. What is the relationship between the following two molecules?

$$\begin{array}{c} \text{CHO} \\ \text{H}_2\text{N}-\!\!\!\!-\!\!\!\!-\text{H} \\ \text{CH}_2\text{OH} \end{array} \qquad \begin{array}{c} \text{CHO} \\ \text{H}-\!\!\!\!-\!\!\!\!-\text{NH}_2 \\ \text{CH}_2\text{OH} \end{array}$$

(A) Regioisomers
(B) Diastereomers
(C) Enantiomers
(D) Identical

82. The following is one of the nucleotides occurring in DNA. What is the hybridization on the two nitrogen atoms indicated in the drawing?

(A) N1 = sp^3 and N2 = sp^2.
(B) N1 = sp^2 and N2 = sp^2.
(C) N1 = sp^3 and N2 = sp^3.
(D) N1 = sp^2 and N2 = sp.

83. The following is a portion of DNA. How many stereoisomers of the molecule are possible?

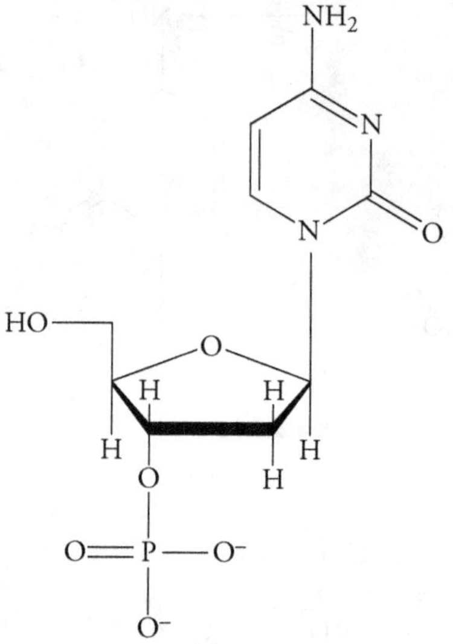

(A) 8
(B) 4
(C) 2
(D) 16

84. Tartaric acid exists in several stereoisomeric forms. One of these forms is:

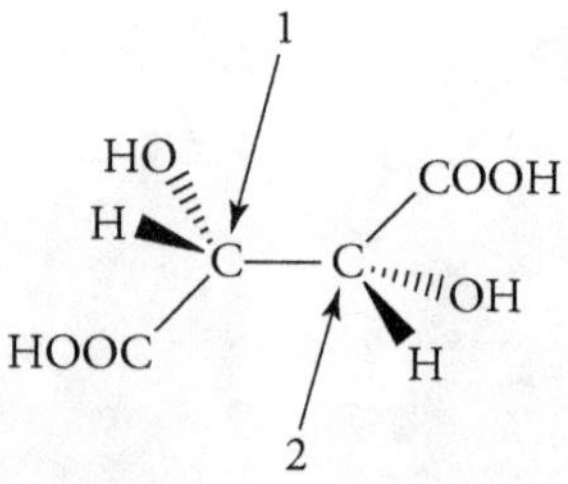

What are the absolute configurations of carbon atoms 1 and 2?
(A) Carbon 1 = S, and carbon 2 = S.
(B) Carbon 1 = R, and carbon 2 = S.
(C) Carbon 1 = R, and carbon 2 = R.
(D) Carbon 1 = S, and carbon 2 = R.

85. Amino acids are classified by their side chains. The side chains may be polar, nonpolar, acidic, or basic. The following molecule is isoleucine. In which category of amino acids is isoleucine?

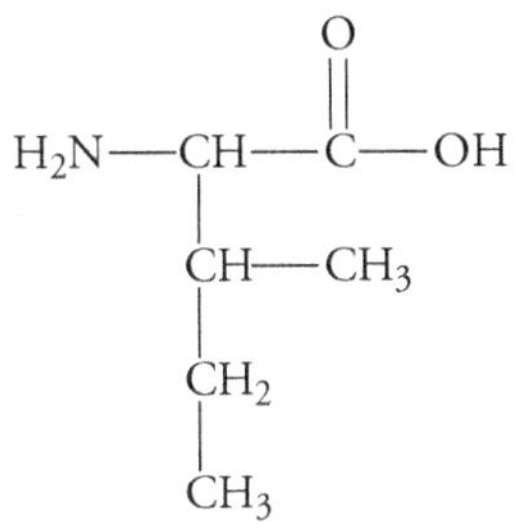

$$H_2N-CH-C(=O)-OH$$
$$|$$
$$CH-CH_3$$
$$|$$
$$CH_2$$
$$|$$
$$CH_3$$

(A) Basic
(B) Nonpolar
(C) Acidic
(D) Polar

86. The following compound has a specific rotation of +12°.

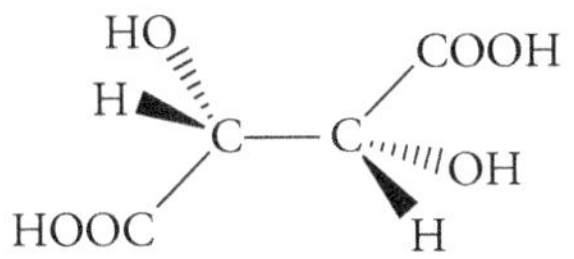

What would be the specific rotation of the compound formed by replacing one of the hydrogen atoms with a fluorine atom?

(A) Impossible to determine from the information given
(B) +12°
(C) −12°
(D) +19°

87. What are the absolute configurations (from left to right) for the chiral centers in the following molecule?

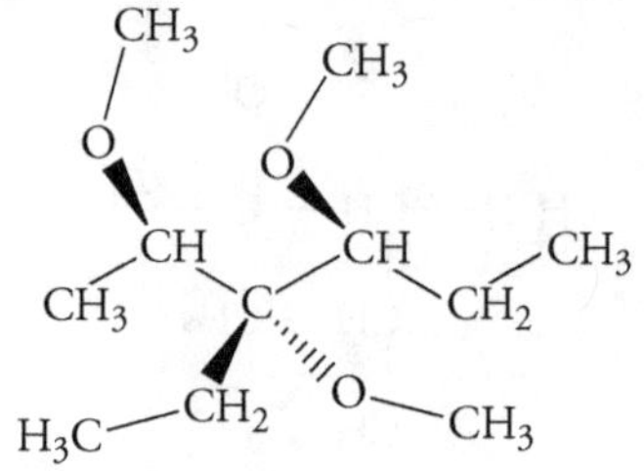

(A) R, R, S
(B) S, S, R
(C) R, R, R
(D) S, S, S

88. How are the following two molecules related?

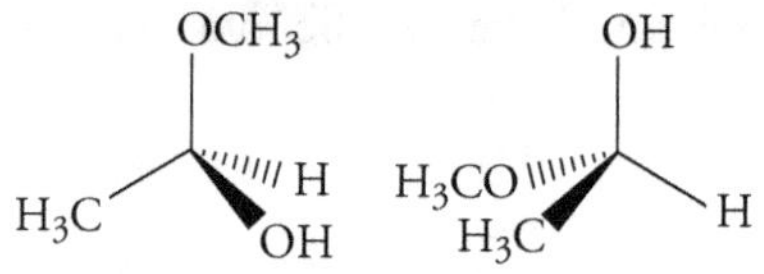

(A) They are structural isomers.
(B) They are enantiomers.
(C) They are diastereomers.
(D) They are the same.

89. The structure of aspartame is:

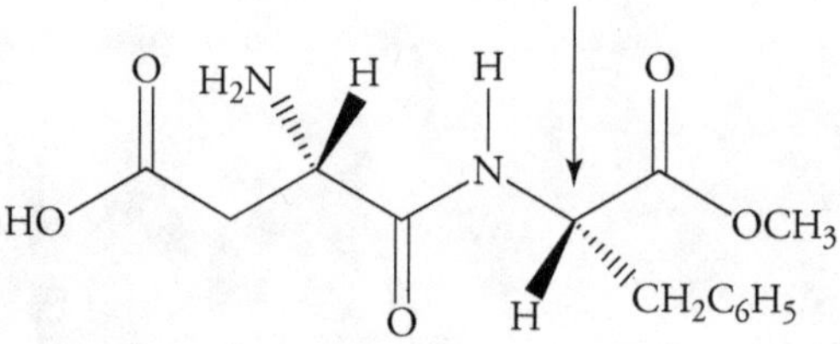

What is the absolute configuration of the indicated carbon in the aspartame?
(A) The carbon is achiral.
(B) The carbon is R.
(C) The carbon is S.
(D) The carbon is meso.

90. What is the relationship between the following two molecules?

   (A) They are identical.
   (B) They are enantiomers.
   (C) They are diastereomers.
   (D) They are structural isomers.

91. The compound 4-nonene exists as cis and trans isomers. Which isomer is more stable?
   (A) The trans isomer is more stable because trans isomers are more reactive.
   (B) The cis isomer is more stable because there is less steric hindrance.
   (C) The cis isomer is more stable because cis isomers are more reactive.
   (D) The trans isomer is more stable because there is less steric hindrance.

92. Many alkenes exist as cis and trans isomers. Is it easy to convert one isomer to the other?
   (A) No, as the conversion requires the breaking of a strong π bond.
   (B) No, as the conversion requires the breaking of a strong σ bond.
   (C) Yes, as the conversion requires the breaking of a weak π bond.
   (D) Yes, as the conversion requires the breaking of a weak σ bond.

93. The compound 1,3,5-hexatriene undergoes catalytic hydrogenation readily. However, the compound 1,3,5-cyclohexatriene does not hydrogenate as readily. Why do the compounds react so differently?
   (A) 1,3,5-hexatriene is very reactive because of destabilization due to resonance.
   (B) 1,3,5-hexatriene is very reactive because of destabilization due to steric hindrance.
   (C) 1,3,5-cyclohexatriene is very unreactive due to resonance.
   (D) 1,3,5-cyclohexatriene is very unreactive due to ring strain.

94. How many stereocenters may a meso compound contain?
    (A) 0
    (B) 1
    (C) 2
    (D) 3

95. What is the term used to describe an optically active compound that rotates plane-polarized light counterclockwise?
    (A) Levorotatory
    (B) Dextrorotatory
    (C) (R)
    (D) (S)

96. The optically active compound (S)-2-methyl-1-butanol has an optical rotation of $[\alpha]_D^{25}$ −5.8°. What is the optical rotation of (R)-2-methyl-1-butanol?
    (A) 0°
    (B) −5.8°
    (C) +5.8°
    (D) 2 (−5.8°)

97. Compounds A and B are enantiomeric alcohols. If compound A boils at 92°C, what is the boiling point of compound B?
    (A) 92°C
    (B) < 92°C
    (C) > 92°C
    (D) Impossible to determine

98. The optical rotation of a certain carboxylic acid is negative, while the optical rotation of its sodium salt is positive. What is one explanation of this observation?
    (A) It is impossible, as the same chiral center is still present.
    (B) There is resonance present in the carboxylate ion.
    (C) There is an increase in the molecular weight when sodium replaces hydrogen.
    (D) The change from a neutral molecule to an ion reverses the optical activity.

99. The following compound was isolated from a reaction:

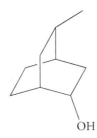

How many stereoisomers are possible for this compound?
(A) 2
(B) 4
(C) 8
(D) 16

100. Carbohydrates can normally exist in the form of several stereoisomers. What is the relationship between the number of carbohydrate stereoisomers, with n = chiral carbon atoms and m = meso forms, to the total number of possible stereoisomers?
(A) 2^n
(B) $2n$
(C) $2n + m$
(D) $2^n + m$

101. Which of the following is NOT true about diastereomers?
(A) They may have different melting points.
(B) The molecular formulas may be different.
(C) The optical rotations may be different.
(D) The gas chromatography retention times may be different.

102. Compounds C and D are enantiomers. Compound D was found to have an R absolute configuration and a positive optical rotation. What are the absolute configuration and optical rotation of Compound C?
(A) R, positive
(B) S, positive
(C) R, negative
(D) S, negative

103. A chemist is investigating a new reaction on cyclohexane. She reasons that there are two possible products. These products are:

Which of the following properties is most likely NOT the same for each of the two possible products?
(A) Optical activity.
(B) Boiling point.
(C) Solubility in water.
(D) All of the properties would be different.

104. What is the stereochemistry of this cyclic amine?

(A) 2R and 3S
(B) 2S and 3R
(C) 2S and 3S
(D) 2R and 3R

105. For which of the following molecules is hydrogen bonding an important factor influencing the physical properties?

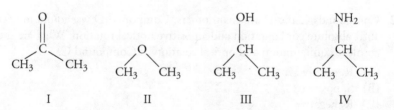

(A) I, II, III, and IV
(B) I, III, and IV
(C) II, III, and IV
(D) III and IV

106. What are the absolute configurations of the following compounds?

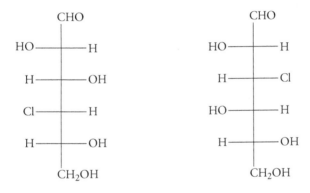

I

II

(A) I is R, and II is S.
(B) I is R, and II is R.
(C) I is S, and II is S.
(D) I is S, and II is R.

107. What is the relationship between the following two molecules?

```
      CHO                          CHO
       |                            |
HO ---+--- H                 HO ---+--- H
       |                            |
 H ---+--- OH                 H ---+--- Cl
       |                            |
Cl ---+--- H                 HO ---+--- H
       |                            |
 H ---+--- OH                 H ---+--- OH
       |                            |
     CH₂OH                        CH₂OH
```

(A) Enantiomers
(B) Regioisomers
(C) Conformers
(D) Identical

108. Which of the following molecules are meso compounds?

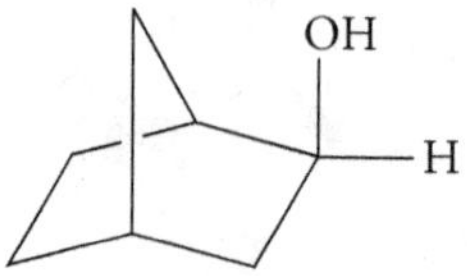

I

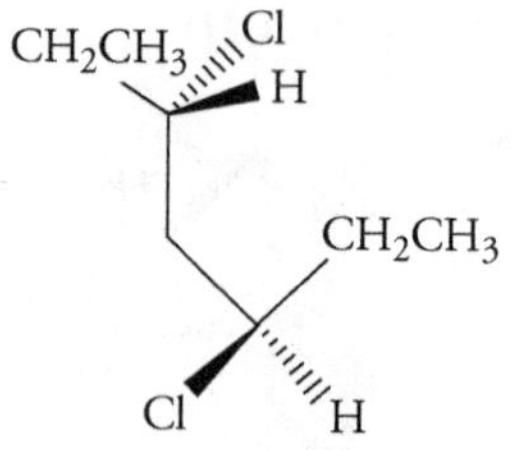

II

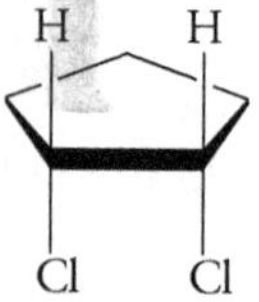

III

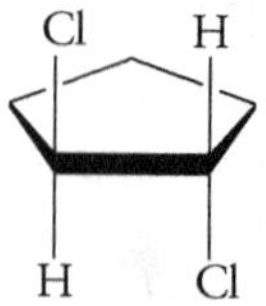

IV

(A) II and III
(B) III and IV
(C) I and III
(D) I and IV

109. How are the boiling points of the following compounds related?

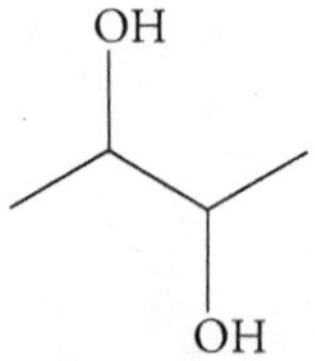

I

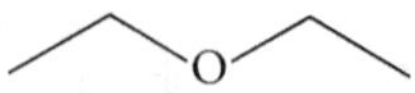

II

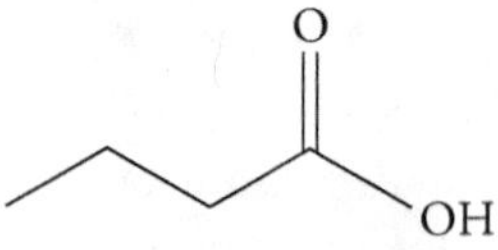

III

(A) III > II > I
(B) II > III > I
(C) III > I > II
(D) I > III > II

110. What is the best explanation of why compound I has a higher boiling point than compound II?

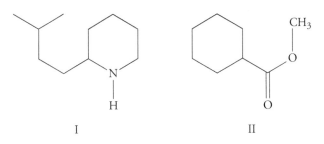

I

II

(A) Compound I can ionize.
(B) Compound II is nonpolar.
(C) Compound I has hydrogen bonding.
(D) Compound II is unstable.

111. Which of the following compounds has an internal plane of symmetry?

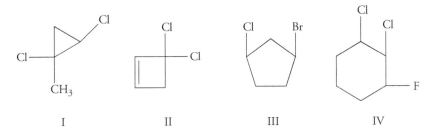

I II III IV

(A) I only
(B) III and IV
(C) II only
(D) None

112. Which of the following molecules is achiral?

(A)

(B)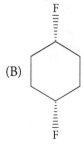

(C)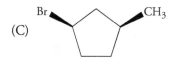

(D)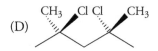

113. An unknown substance has an optical rotation of 0° and a sharp melting point. The unknown substance reacts with the S stereoisomer of another compound to produce two easily separated products with different melting points. Which of the following could describe the unknown substance?

(A) It is a racemic mixture.
(B) It is a cis isomer.
(C) It is an equimolar mixture of two diastereomers.
(D) It has two different compounds present.

114. How many optical isomers does cholesterol have? The structure of cholesterol is:

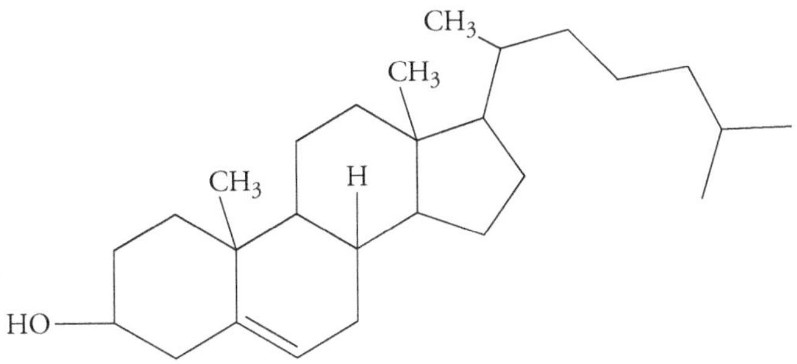

(A) 512
(B) 128
(C) 64
(D) 256

115. The drug thalidomide was introduced in the late 1950s to alleviate the morning sickness that often accompanies pregnancy. Synthesis of the drug resulted in a racemic mixture. One of the enantiomers was the active ingredient for treating morning sickness, while the other enantiomer led to birth defects. The following is one of the enantiomers of thalidomide. What is the absolute configuration of this enantiomer?

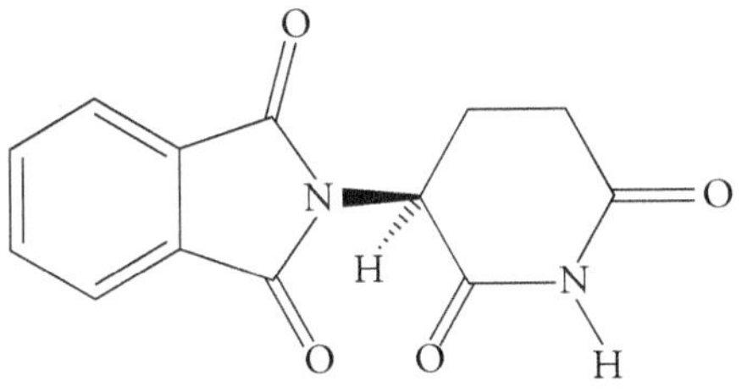

(A) R
(B) S
(C) (+)
(D) (−)

116. Which of the following cyclopropanes is a meso compound?

(A)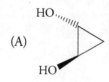

(B)

(C)

(D)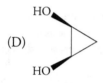

CHAPTER 3

Substitution and Elimination Reactions

Questions 117–125 refer to the passage below.

Among the most important reactions in organic chemistry is substitution reactions, especially S_N2 reactions. Here a nucleophilic group replaces a group on another molecule (the electrophile). Factors that might favor this reaction over an S_N1 reaction include the strength of the nucleophile, the concentration of both the nucleophile and the electrophile, and the solvent chosen.

117. When plotting an S_N2 mechanism on a reaction coordinate graph, which of the following would be at the highest point on the graph?

(A) $\left[RO \overset{\delta+}{-----} \underset{\underset{H}{\overset{H}{|}}}{C} \overset{\delta-}{-----} Br \right]^{\ddagger}$

(B) $\left[RO \overset{\delta-}{-----} \underset{\underset{H}{\overset{H}{|}}}{C} \overset{\delta-}{-----} Br \right]^{\ddagger}$

(C) $\left[\underset{H}{\overset{H}{\diagdown}} \overset{\delta+}{C} \overset{\delta-}{-----} Br \right]^{\ddagger}$

(D) $\left[\underset{H}{\overset{H}{|}} \overset{\oplus}{C} \underset{H}{} \right]^{\ddagger}$

118. Which of the following alkyl halides is most likely to react via an S_N2 mechanism?

(A)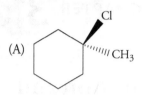

(B)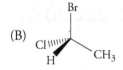

(C)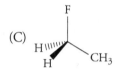

(D)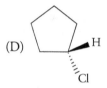

119. Which of the following is NOT a characteristic of an S_N2 mechanism?
   (A) The kinetics are second-order.
   (B) It is a one-step mechanism.
   (C) There is a five-coordinate intermediate.
   (D) There is a carbocation intermediate.

120. The study of an S_N2 reaction on an alkyl halide utilizing hydroxide ion as the nucleophile showed a dependence upon the solvent. In one series of experiments, the changing of the solvent resulted in a slower reaction. What might be the reason why the solvent change lowered the reaction rate?
   (A) The new solvent destabilized the nucleophile.
   (B) The new solvent decreased the activation energy.
   (C) The new solvent increased the value of the rate constant.
   (D) The new solvent solvated the nucleophile more effectively.

121. The main difference between the rate law for an S_N2 mechanism and that for an S_N1 mechanism is that the rate law for an S_N2 process depends only on:
    (A) The concentration of the substrate and the concentration of the nucleophile.
    (B) The concentration of the nucleophile.
    (C) The concentration of the substrate.
    (D) The concentration of the solvent.

122. Which of the following is true concerning an S_N2 mechanism?
    (A) The concentration of the nucleophile is not important.
    (B) It is a two-step process.
    (C) There is retention of configuration.
    (D) Primary alkyl halides react faster than secondary or tertiary alkyl halides.

123. What is the best explanation of why CH_3CH_2Cl reacts via an S_N2 mechanism faster than $CH_3CH_2CH_2CH_2Cl$ does?
    (A) The ethyl group leads to less steric interference.
    (B) Ethyl chloride has a lower boiling point than butyl chloride.
    (C) Ethyl chloride interacts less with the solvent.
    (D) The ethyl group stabilizes the intermediate carbocation better.

124. An organic chemist examined a number of reactions involving an S_N2 mechanism and found a decrease in the rate with increased size of the primary alkyl halide. In addition, there was a decrease in the rate with an increase in the branching on the alkyl halide. What do these facts help to clarify?
    (A) The steric effects on this reaction
    (B) The increased stability of larger alkyls
    (C) The high reactivity of alkyl halides
    (D) The decrease in polarity of larger alkyl halides

125. The monochlorination of methylpropane gives 64% 1-chloro-2-methylpropane and 36% 2-chloro-2-methylpropane. What is the approximate selectivity ratio of 3° to 1° in the reaction?
    (A) 64:36
    (B) 36:64
    (C) 1:9
    (D) 5:1

126. An organic chemist uses the following reaction to produce the starting material for another reaction. The reaction takes place in dry ether and gives one organic product.

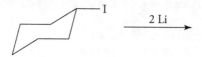

What is the major product of this reaction?

(A)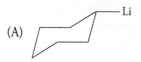

(B)

(C)

(D)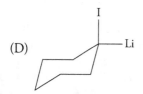

127. It is possible to form an ether through the reaction of an alkoxide with a primary alkyl halide. What is the function of the alkoxide in this reaction?
   (A) It is an electrophile.
   (B) It is a nucleophile.
   (C) It is an acid.
   (D) It is a catalyst.

128. How many products form in the monochlorination of 2,4-dimethylpentane?
   (A) 12
   (B) 8
   (C) 4
   (D) 1

129. An organic chemist investigates the following reaction:

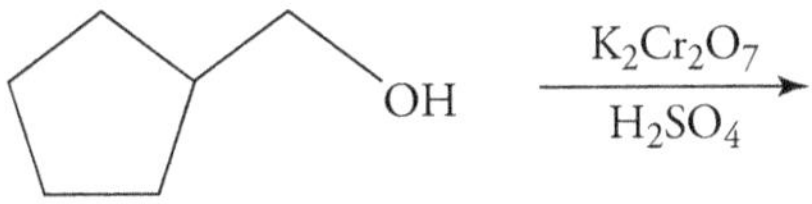

[cyclopentyl-CH2-OH] $\xrightarrow{K_2Cr_2O_7 / H_2SO_4}$

What is the major product?

(A) 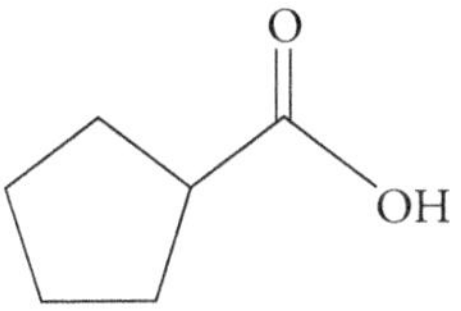 cyclopentyl-C(=O)-OH

(B) 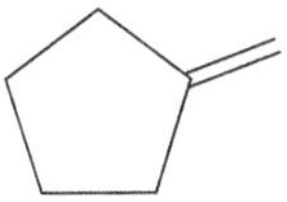 methylenecyclopentane

(C) cyclopentyl-CHO

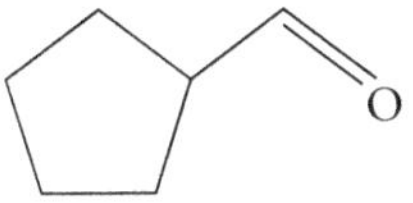

(D) 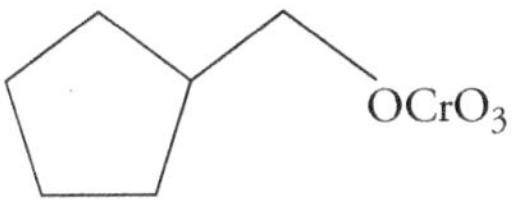 cyclopentyl-CH2-OCrO3

130. An organic chemist investigates the following reaction:

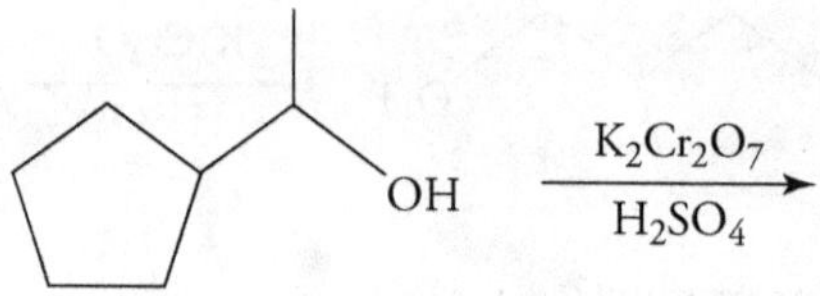

cyclopentyl-CH(OH)-CH3 $\xrightarrow{K_2Cr_2O_7 / H_2SO_4}$

What is the major product?

(A) 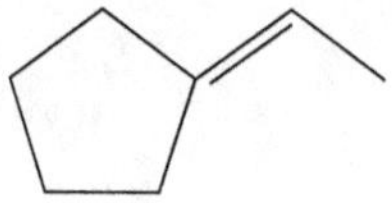 cyclopentylidene-CHCH3 (alkene)

(B) 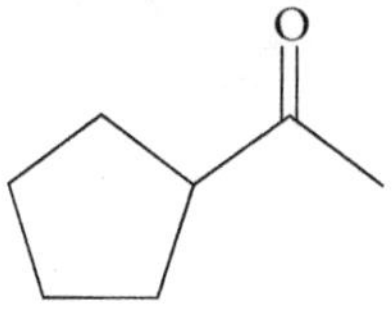 cyclopentyl-C(=O)-CH3

(C) cyclopentyl-CHO

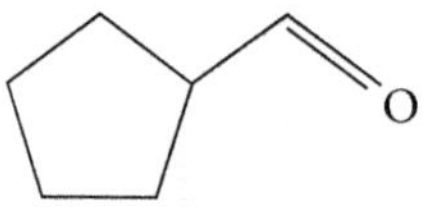

(D) 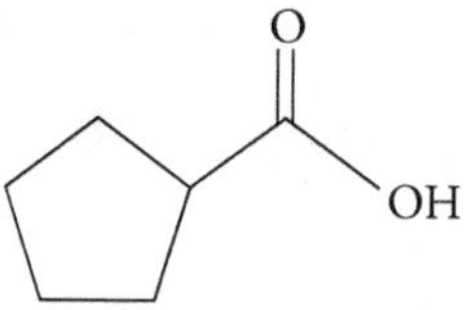 cyclopentyl-C(=O)-OH

131. An organic chemist investigates the following reaction:

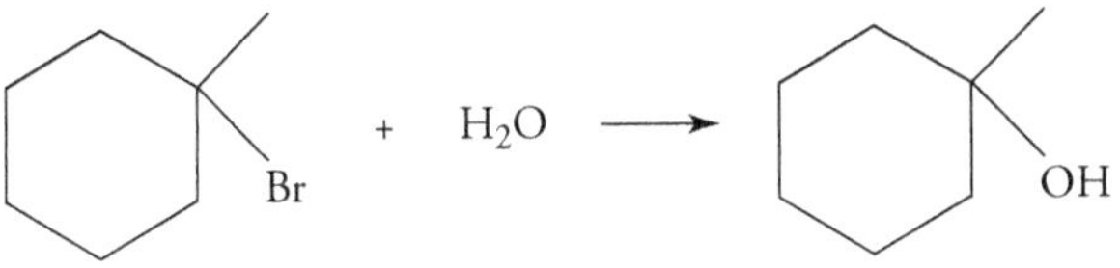

[Structure: 1-bromo-1-methylcyclohexane] + H_2O → [Structure: 1-methylcyclohexan-1-ol]

Which of the following is the most likely reaction intermediate?

(A) 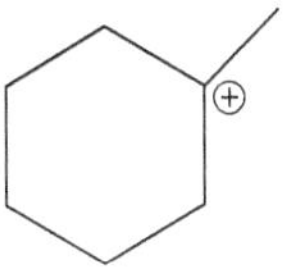 [Cyclohexane with methyl group and carbocation (+)]

(B) 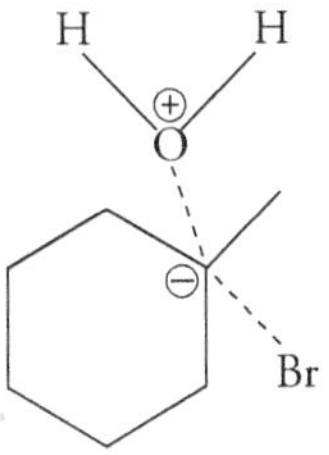 [Cyclohexane with Br leaving and H₂O⁺ attacking, showing two H's on O]

(C) 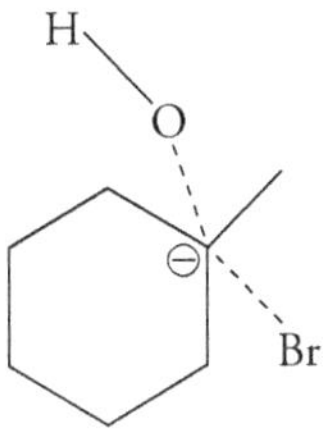 [Cyclohexane with Br leaving and HO⁻ attacking, showing one H on O]

(D) 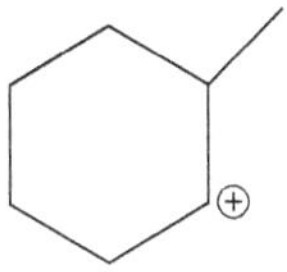 [Cyclohexane with methyl group and carbocation (+) on ring carbon]

132. Which of the following is the best leaving group?
 (A) OH⁻
 (B) CH₃O⁻
 (C) NH₃
 (D) NH₂⁻

133. Which of the following leads to the best leaving group?
    (A) C–N=NH
    (B) C–NH-NH$_2$
    (C) C–NH$_2$
    (D) C–$^+$N≡N

134. Which of the following illustrates a propagation step in a free radical chain reaction?

(A) :Cl· + (CH$_3$)$_2$CH–CH$_3$ ⟶ (CH$_3$)$_2$C·–CH$_3$ + HCl

(B) 2 :Cl· ⟶ :Cl—Cl:

(C) (CH$_3$)$_3$C· + :Cl· ⟶ (CH$_3$)$_3$C–Cl

(D) :Cl—Cl: $\xrightarrow{h\nu}$ 2 ·Cl:

135. Which of the following free radicals is the most stable?

(A) benzene ring with –CH(CH₃)– radical substituent (radical on the benzylic carbon)

(B) benzene ring with isopropyl substituent and radical on a ring carbon (ortho position)

(C) benzene ring with isopropyl substituent where the radical is on the CH of the isopropyl group

(D) benzene ring with isopropyl substituent and radical on a ring carbon (meta position)

136. An organic chemist investigates the following reaction:

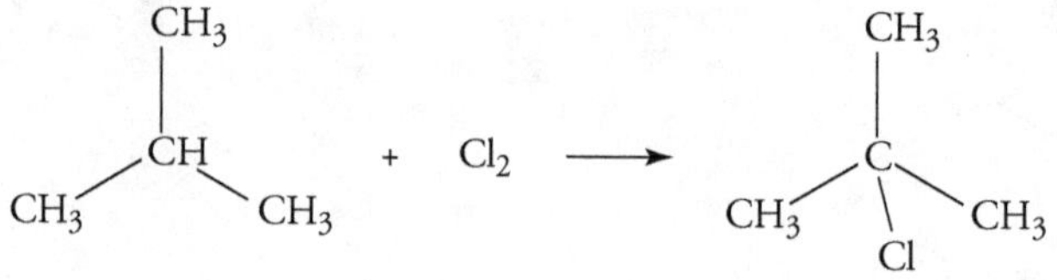

(CH₃)₂CH—CH₃ + Cl₂ ⟶ (CH₃)₃C—Cl

Which of the following intermediates is most likely to be important to the reaction?

(A) Cl----C(CH₃)₂----H with CH₃

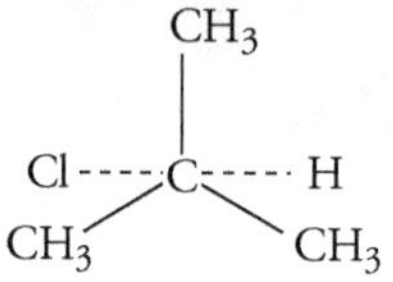

(B) (CH₃)₂C(H)—ĊH₂

(C) (CH₃)₃C⁺

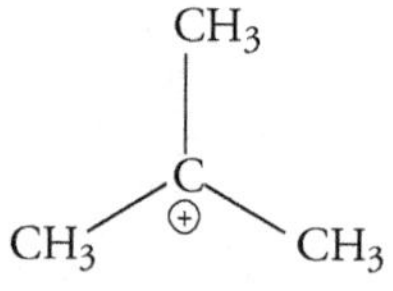

(D) (CH₃)₃Ċ

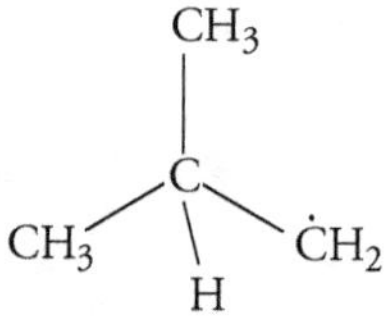

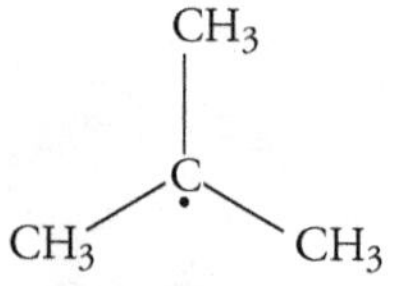

137. Which of the following alcohols CANNOT be oxidized to a carbonyl compound by acidic $K_2Cr_2O_7$?

(A) CH_3OH

(B) ![isopropanol structure]

(C) ![tert-butanol structure]

(D) ![1-phenylethanol structure]

138. Which of the following alcohols are most likely to react with HBr through an S_N1 mechanism?

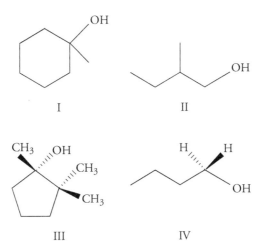

(A) I and III
(B) II and IV
(C) III and IV
(D) II and III

139. Which of the following is the major product from the dehydration of 2-methyl-1-pentanol?

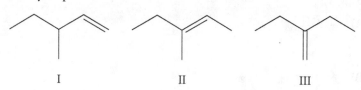

(A) I
(B) II
(C) III
(D) All form in equal amounts.

140. Which of the following is the major product formed by the free radical monobromination of isopentane?

(A)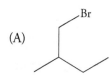

(B)

(C)

(D)

141. Why is structure II more stable than structure I?

$$\text{Structure I: } CH_3\text{-}C(CH_3)_2\text{ (axial)}, CH_2CH_3 \text{ (axial)}$$
$$\text{Structure II: } CH_3\text{-}C(CH_3)_2\text{ (equatorial)}, CH_2CH_3 \text{ (equatorial)}$$

(A) The two groups can be axial, so there is less repulsion.
(B) The two groups can be equatorial, so there is less repulsion.
(C) The t-butyl group can be axial, so there is less repulsion with the equatorial ethyl group.
(D) The t-butyl group can be equatorial, so there is less repulsion with the axial ethyl group.

142. Which of the following molecules is the better nucleophile, AsH_2^- or PH_2^-?

(A) AsH_2^-
(B) PH_2^-
(C) It is not possible to tell.
(D) They are the same.

143. The acetate ion is a stronger base than the trichloroacetate ion. Why is it a stronger base?

(A) Inductive effects
(B) Higher polarity
(C) Resonance effects
(D) Steric hindrance

144. Place the following carbocations in order of increasing stability.

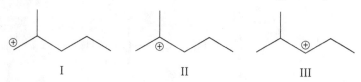

(A) I < III < II
(B) II < I < III
(C) II < III < I
(D) I < II < III

145. The reaction of a certain alkyl bromide utilizes a carbocation intermediate. The carbon in the carbocation has what type of hybridization?
(A) sp^3
(B) sp^2
(C) sp
(D) p^3

146. What is/are the product(s) of an S_N1 reaction by an achiral molecule on the chiral center of another molecule?
(A) It produces a racemic mixture.
(B) The product has inversion of configuration.
(C) The product retains the original configuration.
(D) The product is achiral.

147. Which of the following is NOT a characteristic of an E1 mechanism?
(A) The kinetics are first-order.
(B) There is an attack by a nucleophile on a carbocation.
(C) A π bond forms.
(D) It is a two-step mechanism.

148. Which of the following is NOT part of an E2 mechanism?
(A) There are two carbon atoms changing from sp^3 to sp^2 hybridization.
(B) There is the formation of a π bond.
(C) There is the formation of a carbocation intermediate.
(D) There is the loss of a leaving group.

149. What type of reaction is the chlorination of an alkane?
  (A) It is a free radical reaction.
  (B) It is an S_N1 reaction.
  (C) It is an S_N2 reaction.
  (D) It is an E2 reaction.

150. The study of a series of alkyl halides with the general formula RR'R"C-X, where the different Rs may or may not be the same and X is Cl, Br, or I, found that the rate of the S_N1 reaction decreased with an increase in the average size of the R groups. Why did the rate decrease?
  (A) There is an increase in the stability of the product.
  (B) There is a loss of optical activity.
  (C) There is an increase in the activation energy.
  (D) The mechanism shifts from S_N1 to S_N2.

151. Which of the following is applicable to the reaction of methyl bromide under nucleophilic substitution conditions?
  (A) The kinetics are second-order.
  (B) A racemic mixture will form.
  (C) There is retention of configuration.
  (D) The kinetics are first-order.

152. Which of the following illustrates a termination step in a free radical chain reaction?

(A) :Cl· + (CH₃)₃CH → (CH₃)₃C· + HCl

(B) (CH₃)₃C· + :Cl· → (CH₃)₃C-Cl

(C) :Cl—Cl: + :Cl· → :Cl· + :Cl—Cl:

(D) :Cl—Cl: —hv→ 2 ·Cl:

153. Which of the following is NOT characteristic of an S_N2 mechanism?
    (A) The reactivity sequence decreases in the order 1° > 2° > 3°.
    (B) There is inversion of configuration.
    (C) There is no rearrangement.
    (D) The kinetics are first-order.

154. An organic chemist is investigating the reaction of ethyl iodide with the hydroxide ion. What was the result of doubling the concentration of hydroxide ion upon the rate of the reaction?
    (A) The rate quadruples.
    (B) The rate doubles.
    (C) The rate remains the same.
    (D) The rate triples.

155. Which of the following best describes the free radical chlorination of an alkane?
    (A) The reaction is nonregioselective, leading to a number of isomers due to inductive and statistical factors.
    (B) The reaction is nonregioselective, leading to the most stable isomer due to inductive and statistical factors.
    (C) The reaction is stereoselective, producing primarily one enantiomer.
    (D) The reaction is nonstereoselective, producing a racemic mixture of the two enantiomers.

156. What is the name of the first step in a free radical halogenation mechanism?
    (A) Initiation
    (B) Commencement
    (C) Instigation
    (D) Propagation

157. The R form of the following compound will react with a mixture of NaI in acetone:

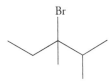

What is the major product of the reaction?
(A) A racemic mixture of 3-iodo-2,3-dimethylpentane
(B) (R)-3-iodo-2,3-dimethylpentane
(C) (S)-3-iodo-2,3-dimethylpentane
(D) 2,3,3-trimethylpentane

158. What are the products of the room-temperature reaction of one equivalent of hydrogen chloride gas and one equivalent of 1,3-butadiene?
(A) 3-chloro-1-butene plus 1-chloro-2-butene
(B) 2-chloro-1-butene plus 1-chloro-2-butene
(C) 2-chloro-1-butene plus 2-chloro-2-butene
(D) 1-chloro-1-butene plus 1-chloro-2-butene

159. Which molecule is the better nucleophile, OH^- or NH_2^- ?
(A) OH^-
(B) NH_2^-
(C) It is not possible to tell.
(D) They are the same.

160. What is the product of the reaction of 2-hexene with $CHCl_3$ in the presence of potassium t-butoxide?

(A) 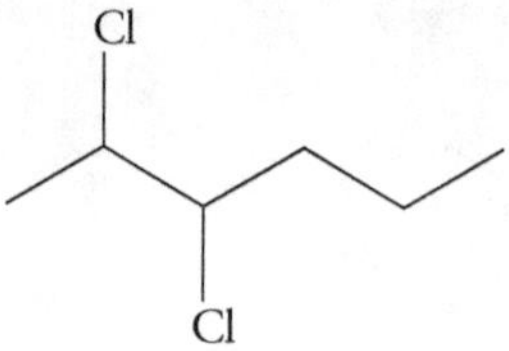 [structure with two Cl groups on adjacent carbons of a hexane chain]

(B) 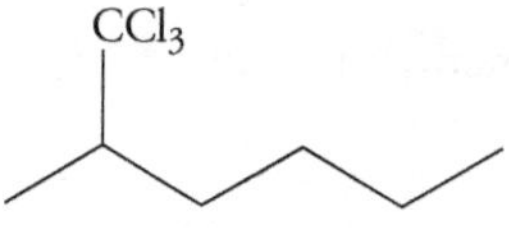 [structure with CCl_3 group on hexane chain]

(C) 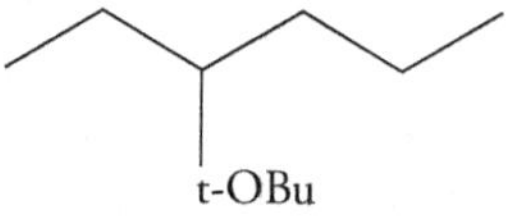 [structure with t-OBu group on hexane chain]

(D) 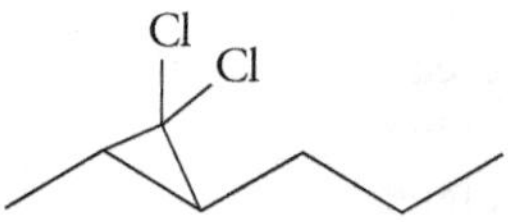 [cyclopropane ring with two Cl substituents fused to propyl chain]

161. What is the purpose of the propoxide ion in the following reaction?

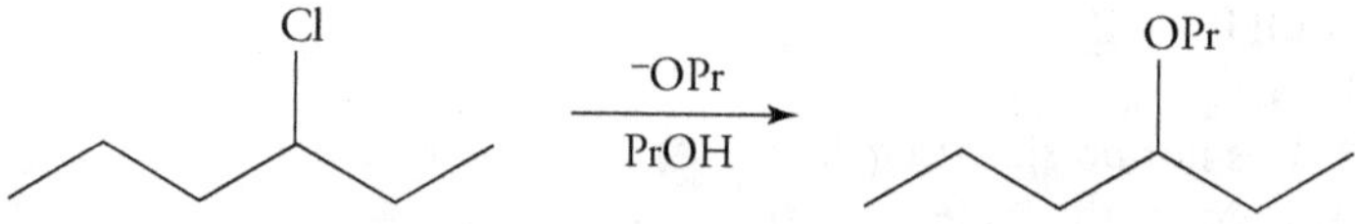

[structure with Cl] $\xrightarrow[PrOH]{^-OPr}$ [structure with OPr]

(A) This is an S_N2 reaction, and the propoxide ion is the nucleophile.
(B) This is an S_N2 reaction, and the propoxide ion is the electrophile.
(C) This is an S_N1 reaction, and the propoxide ion is the base.
(D) This is an S_N1 reaction, and the propoxide ion is the electrophile.

162. The free radical monobromination of methylcyclopentane forms only the following product:

What is the best explanation of this observation?
(A) There is stabilization of the intermediate through resonance.
(B) There is an attack at the least sterically hindered site.
(C) There is stabilization of the intermediate through inductive effects.
(D) There is an attack on the site that is the most δ+.

163. What occurs in the propagation steps of all free radical addition reactions?
(A) A molecule reacts to produce two free radicals.
(B) A free radical reacts to produce another free radical.
(C) Two free radicals react to produce a molecule.
(D) A free radical reacts to produce a carbocation.

164. In the presence of sodium ethoxide, 3-chloro-3-ethylpentane reacts primarily by an S_N1 mechanism instead of an S_N2 mechanism. How will the rate of the reaction change if the sodium ethoxide concentration is doubled?
(A) There is a decrease in the rate of reaction.
(B) There is an increase in the rate of reaction.
(C) There is no change in the rate of reaction.
(D) There is a change from an S_N1 to an S_N2 mechanism.

165. The carbocation formed by the reaction of an unknown alkyl bromide is more stable than the carbocation formed by the reaction of neopentyl bromide. Which of the following might the alkyl bromide be?
(A) Isopropyl bromide
(B) Ethyl bromide
(C) Methyl bromide
(D) Propyl bromide

166. Which of the following compounds is the most likely to react via an S_N1 mechanism?
(A) CH_3CH_2I
(B) $(CH_3)_3CCH_2I$
(C) $(CH_3)_2CHI$
(D) $(CH_3)_3CI$

167. Place the following free radicals in order of decreasing stability.

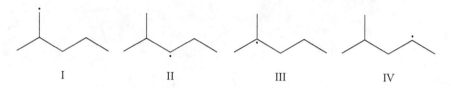

(A) III > II > IV > I
(B) III > II ≈ IV > I
(C) I > IV > II > III
(D) I > III > IV > II

168. Phenol is a weak acid. Which of the following benzene-ring substituents will increase its strength as an acid?

$-CN$ $-CH_2CH_3$ $-CCl_3$
I   II   III

(A) I and III
(B) I and II
(C) II and III
(D) I, II, and III

169. Which of the following is the most stable conformation of 1,2-ethanediol? (The remaining hydrogen atoms have been removed for clarity.)

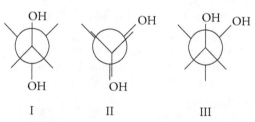

(A) I
(B) II
(C) III
(D) All are equal.

170. Which of the following are the products formed by the free radical chlorination of isopentane?

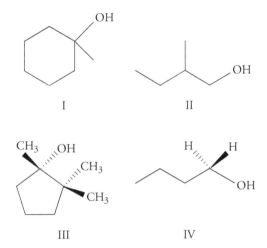

(A) I, II, III, and IV
(B) II and IV
(C) I and III
(D) I, II, and III

171. Which of the following alcohols are most likely to react with HBr through an S_N2 mechanism?

(A) III and IV
(B) I and III
(C) II and IV
(D) II and III

172. What is the major product of the following reaction?

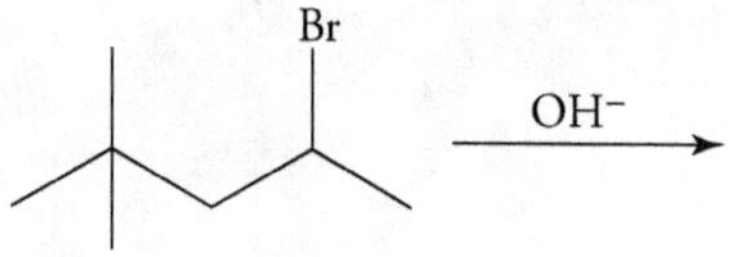

(A)

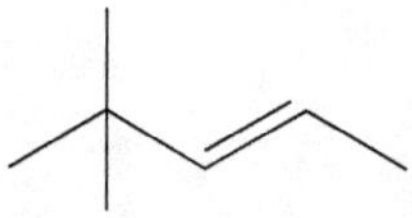

(B)

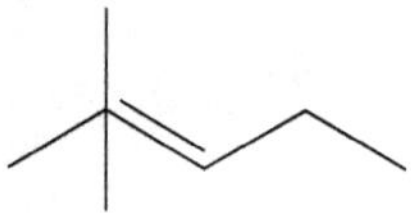

(C)

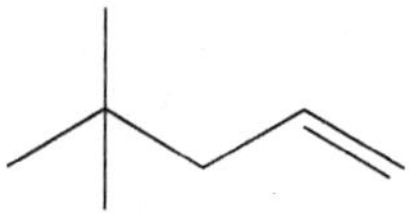

(D) No reaction

173. An organic chemist investigates the following reaction in the presence of light:

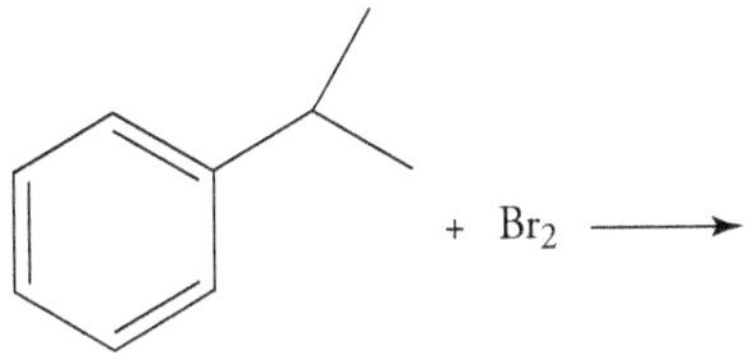

isopropylbenzene + Br_2 ⟶

Which of the following will be the major product?

(A) 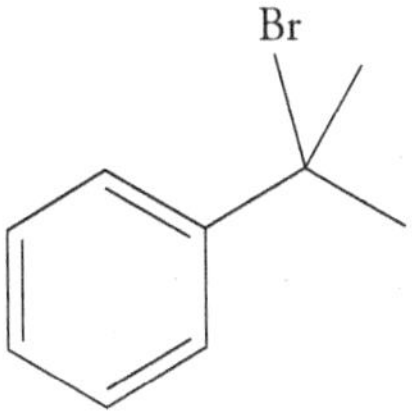 2-bromo-2-phenylpropane (Br on benzylic carbon)

(B) 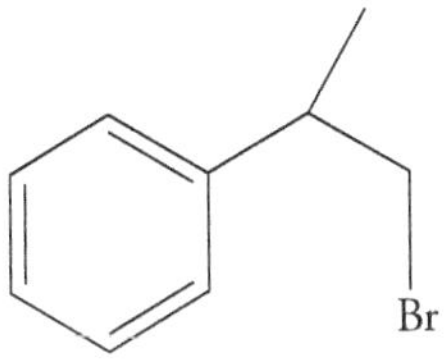 1-bromo-2-phenylpropane (Br on primary carbon)

(C) 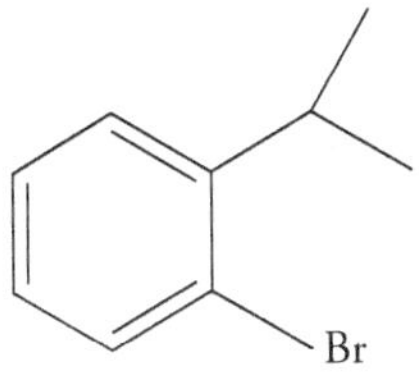 ortho-bromo isopropylbenzene

(D) 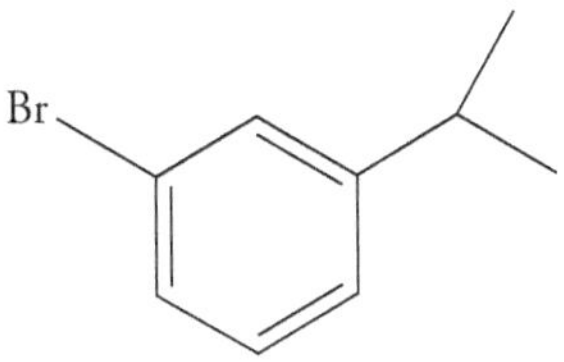 meta-bromo isopropylbenzene

174. An organic chemist investigates the following reaction:

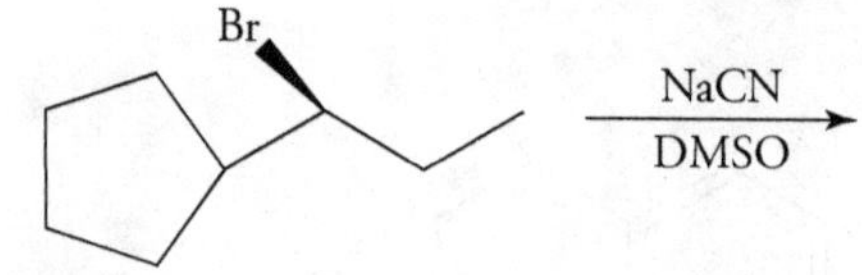

Which of the following will be the major product?

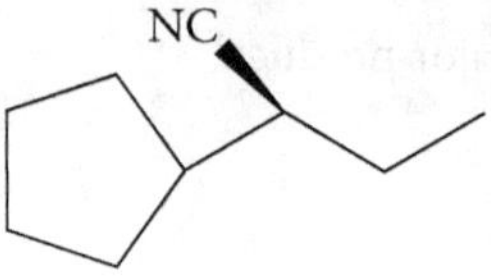

I

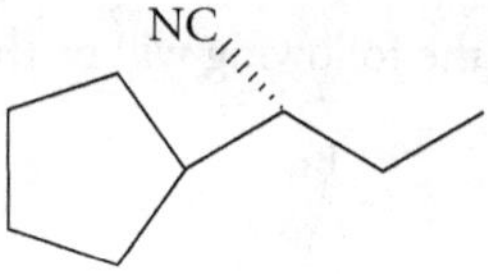

II

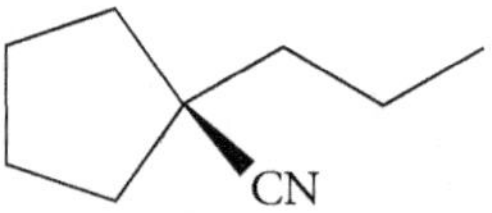

III

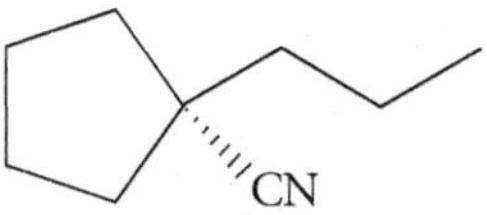

IV

(A) I only
(B) II only
(C) I and II
(D) III and IV

175. An organic chemist investigates the following reaction:

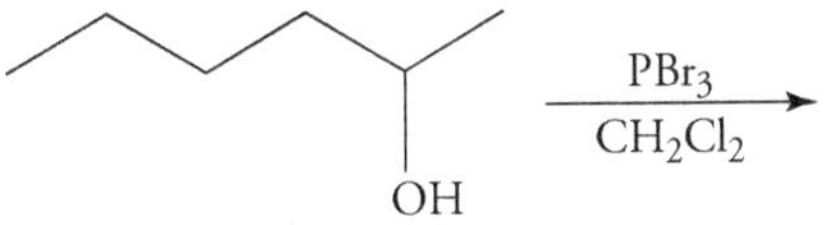

CH₃CH₂CH₂CH(OH)CH₃ → (PBr₃, CH₂Cl₂)

Which of the following will be the major product?

(A) 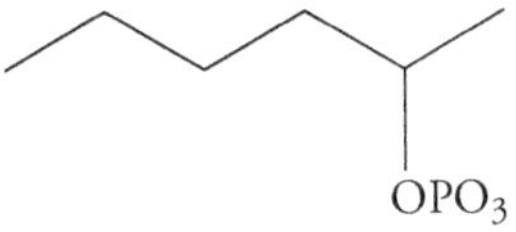 structure with OPO₃

(B) 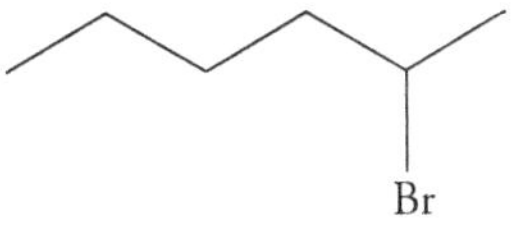 structure with Br (2-bromopentane)

(C)  pentene

(D) 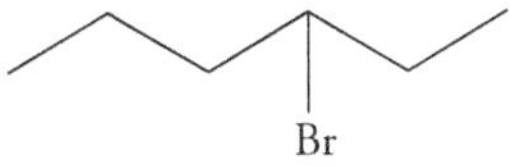 structure with Br on C3

176. Which of the following nitrogen-containing species will NOT serve as a nucleophile?

(A) 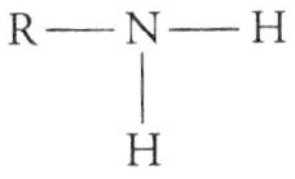 R—NH—H (R-NH₂, with H below)

(B) 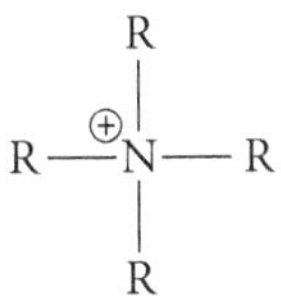 R—N⁺(R)(R)—R (quaternary ammonium)

(C) 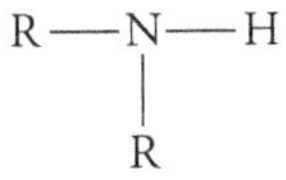 R—NH—H with R below

(D) 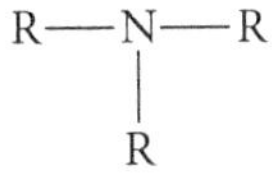 R—N(R)—R with R below

177. The reaction of t-butyl iodide with formate ion occurs more rapidly in aqueous solution than in DMF solutions. Why is this the case?
   (A) DMF reacts with the formate ion.
   (B) Water is smaller, which leads to less steric hindrance.
   (C) Water stabilizes the intermediate carbocation better.
   (D) DMF forms a less reactive intermediate.

178. A chemist is investigating the following type of reaction using primary and secondary alcohols. After mixing the reactants, the mixture is refluxed.

$$ROH + SOCl_2 \rightarrow$$

Which of the following compounds is the most likely product?

(A) R—O—S(=O)(=O)—OH

(B) R—S—C(=O)—O

(C) R—Cl

(D) R—S—H

179. It is possible to dehydrate an alcohol by heating the alcohol with phosphoric acid. Which of the following alcohols would undergo dehydration most rapidly?

(A) cyclopentyl-CH(OH)-CH3

(B) cyclopentyl-CH2-OH

(C) cyclopentyl-CH2-CH2-OH

(D) 1-methylcyclopentyl with OH and CH3 on same carbon

CHAPTER 4

Electrophilic Addition Reactions

Questions 180–183 refer to the passage below.

Another important reaction in organic chemistry is addition reaction. Here an unsaturated compound (containing double or triple bonds) reacts with another group that adds across that double or triple bond to give a saturated compound. If the unsaturated compound is symmetrical, such as ethylene, then addition of HBr will yield a single product. However, if the unsaturated compound is not symmetrical, then there is a possibility of more than one compound being formed. The Markovnikov rule allows us to predict which product will predominate. It can be summarized as, "Them that has, gets."

180. The following molecule is the starting material for the Markovnikov addition of water. What is the best description of the intermediate formed in the first step of the reaction mechanism?

    (A) A tertiary carbonium ion
    (B) A secondary carbonium ion
    (C) A tertiary carbanion
    (D) A secondary carbanion

181. Alkenes will undergo addition reactions with compounds such as HBr. What best describes the mechanism of this reaction in the presence of peroxides?
    (A) An sp intermediate forms.
    (B) There is no initiation step.
    (C) The reaction will form the Markovnikov product.
    (D) The reaction will form the anti-Markovnikov product.

182. The reaction of HCl with Z-3-methyl-3-heptene is most likely:
    (A) Syn and anti addition to give the Markovnikov products.
    (B) Anti addition to give the Markovnikov product.
    (C) Syn and anti addition to give the anti-Markovnikov products.
    (D) Anti addition to give the anti-Markovnikov product.

183. What type of intermediate should form during the Markovnikov addition of HBr to the following compound?

    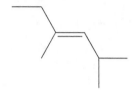

    (A) A secondary carbanion
    (B) A secondary carbonium ion
    (C) A tertiary carbanion
    (D) A tertiary carbonium ion

184. Sulfonation of toluene produces p-toluenesulfonic acid and o-toluenesulfonic acid. What is the relationship between these two products?
    (A) Anomers
    (B) Conformers
    (C) Constitutional isomers
    (D) Diastereomers

185. The following illustrates a generic Diels-Alder reaction:

    What is the net change in the number of σ and π bonds?
    (A) −2 π and +2 σ
    (B) −2 π and +3 σ
    (C) −3 π and +3 σ
    (D) −3 π and +2 σ

186. Hydrogen bromide will react with $CH_3\text{-}CH=CH\text{-}CH=CH\text{-}CH_2CH_2CH_3$. What is the reaction intermediate?
    (A) A conjugate base
    (B) A carbanion
    (C) A free radical
    (D) A carbonium ion

187. Hydrogen halides will add to an alkene. Which halide reacts most rapidly with an alkene in a nonpolar, aprotic solvent?
    (A) HBr
    (B) HI
    (C) HCl
    (D) HF

188. An organic chemist investigated the following reaction and isolated two products:

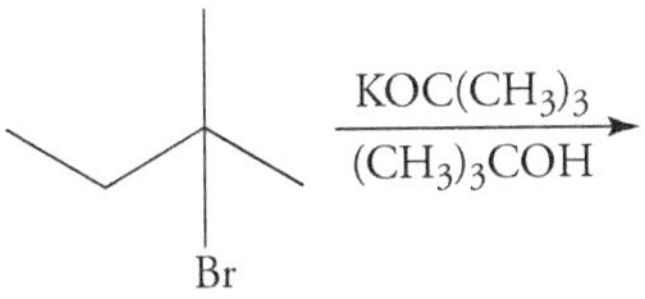

$$\text{(2-bromo-2-methylbutane)} \xrightarrow[\text{(CH}_3)_3\text{COH}]{\text{KOC(CH}_3)_3}$$

What was the structure of the major product?

(A) 2-methyl-1-butene

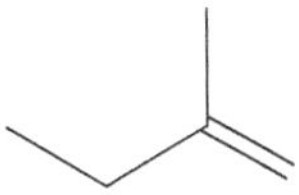

(B) 2-methyl-2-butene

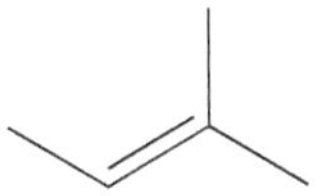

(C) 2-methyl-2-butanol (OH)

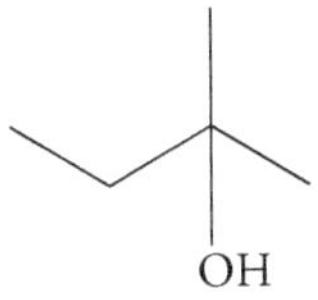

(D) 2-methyl-2-(tert-butoxy)butane (OC(CH$_3$)$_3$)

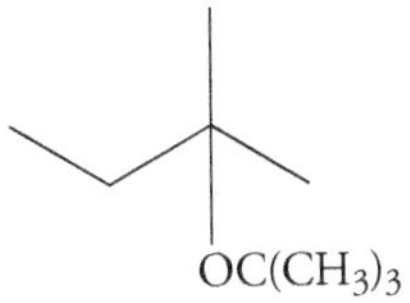

189. Rank the following in increasing order of reactivity rate with $Br_2/FeBr_3$.

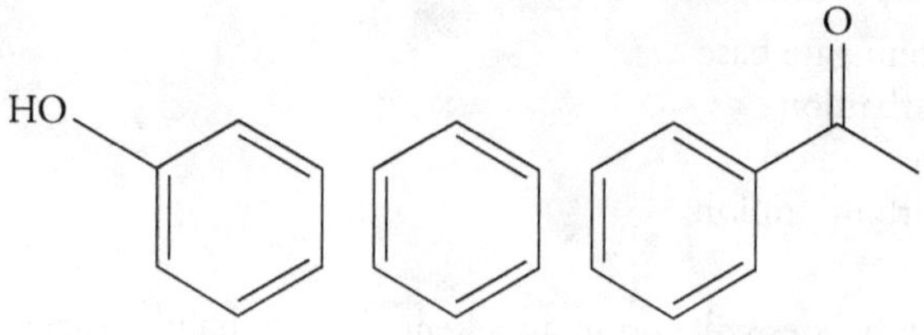

I: HO-phenyl II: benzene III: acetophenyl

(A) II < I < III
(B) I < II < III
(C) II < III < I
(D) III < II < I

190. The following reaction produces one major product:

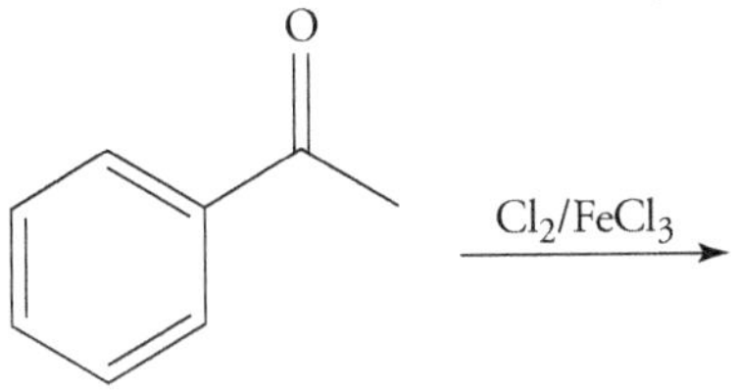

acetophenone $\xrightarrow{Cl_2/FeCl_3}$

What is the structure of the major product?

(A) 2-chloroacetophenone (ortho-Cl)

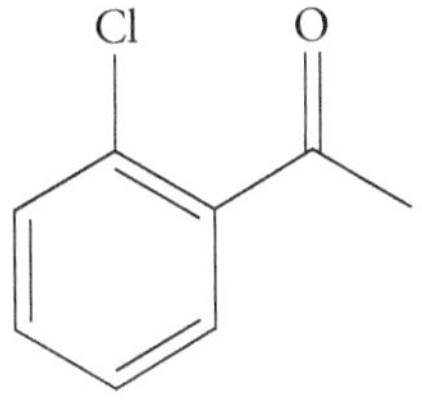

(B) 3-chloroacetophenone (meta-Cl, Cl upper left)

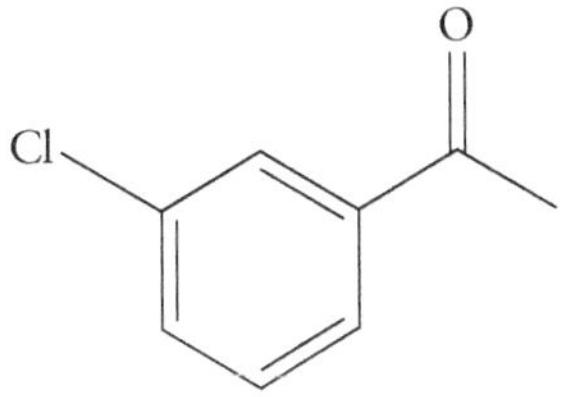

(C) 3-chloroacetophenone (meta-Cl, Cl lower)

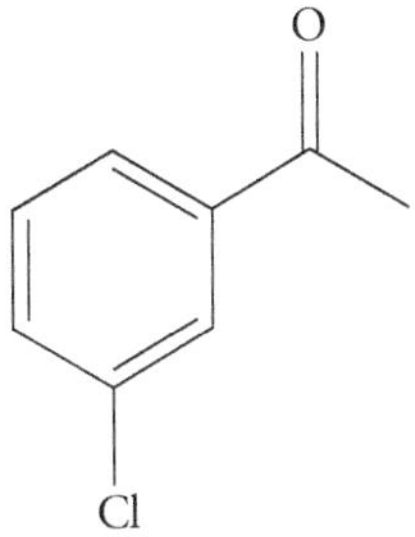

(D) α-chloroacetophenone (Cl on the methyl carbon)

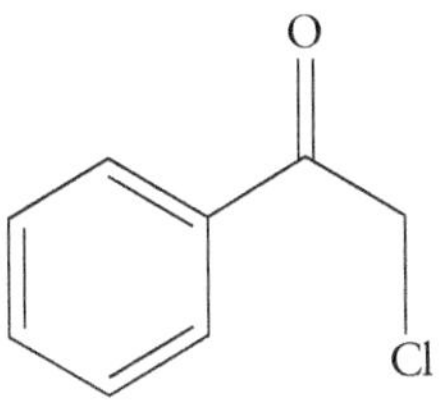

191. It is possible to produce mostly one alkyl halide via the following reaction:

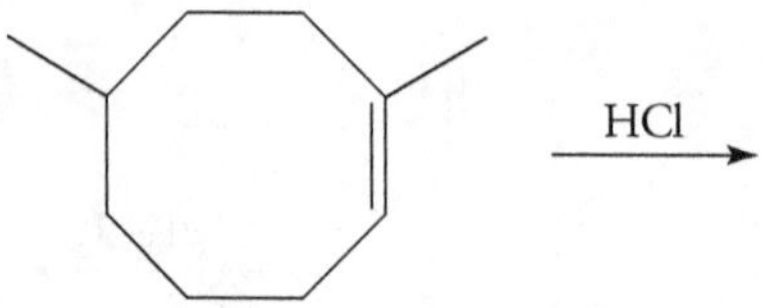

What is the structure of the major product?

(A)

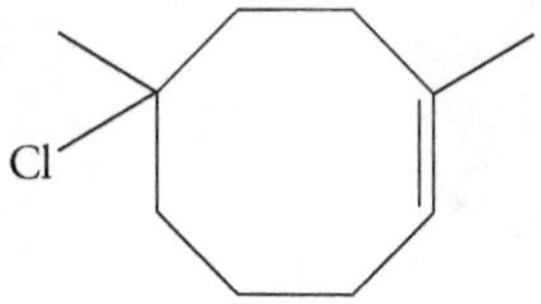

(B)

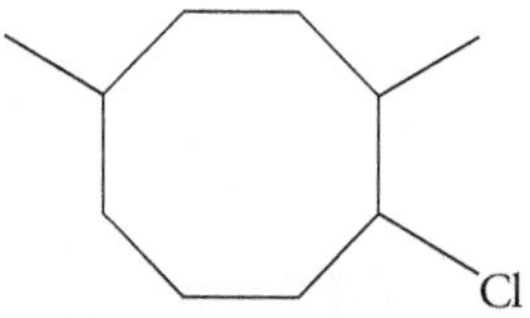

(C)

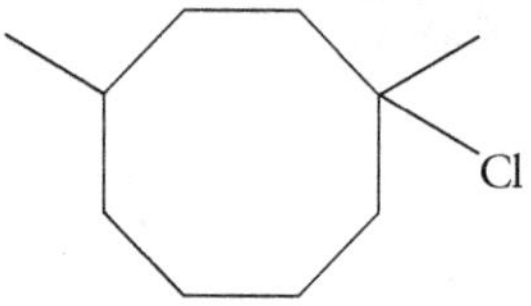

(D)

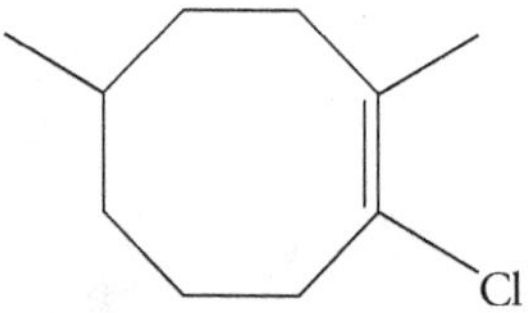

192. What is the major product of the reaction of bromine in carbon tetrachloride with cyclohexene?
    (A) Racemic 1,2-dibromocyclohexane
    (B) Meso 1,2-dibromocyclohexane
    (C) 1,1-dibromocyclohexane
    (D) 1-bromocyclohexane

193. Which of the following is the major product of the reaction of trans-2-butene with meta-chloroperoxybenzoic acid (mCPBA) in methylene chloride?

(A)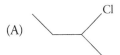

(B)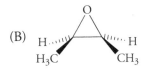

(C)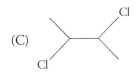

(D)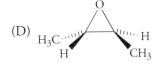

194. A chemist conducts the following two-step chemical procedure:

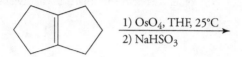

What is the major product of the reaction?

(A)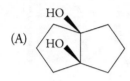

(B)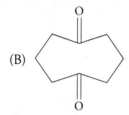

(C)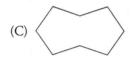

(D)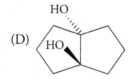

195. A chemist conducts the following two-step chemical procedure:

1) B_2H_6/THF
2) H_2O_2/NaOH/H_2O

What is the major product of the reaction?

(A) cyclopentane with OH on C1 and OH on C2 (both substituents, with methyl)

(B) cyclopentane with methyl and OH geminal

(C) cyclopentane with methyl and OH on same carbon

(D) methylcyclopentane

196. A chemist conducts the following two-step chemical procedure:

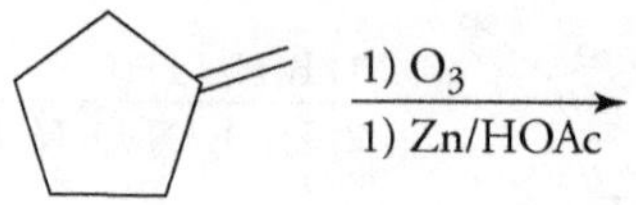

methylenecyclopentane → 1) O_3 2) Zn/HOAc

What is one of the major product(s) of the reaction?

(A) 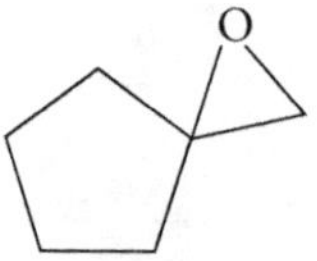 spiro epoxide (cyclopentane with epoxide)

(B) 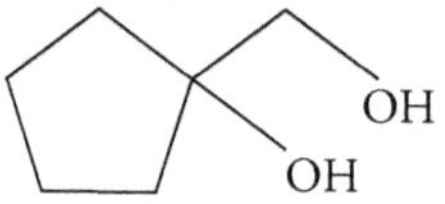 cyclopentane with C(CH$_2$OH)(OH)

(C) cyclopentanone with =O exocyclic

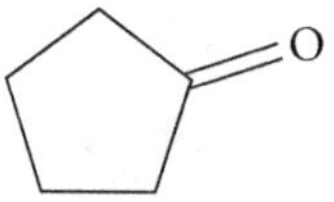

(D) 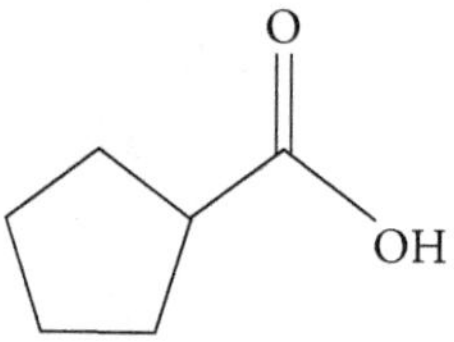 cyclopentane-COOH

197. The following reaction produces one major product:

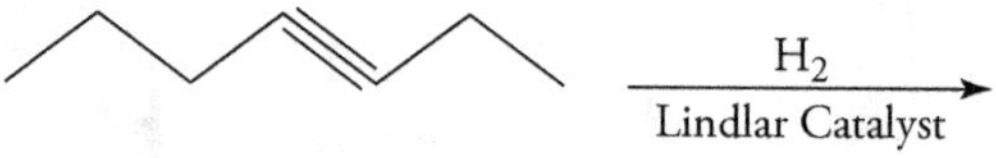

CH$_3$CH$_2$CH=CHCH$_2$CH$_3$ (actually alkyne shown) $\xrightarrow{\text{H}_2, \text{ Lindlar Catalyst}}$

What is the major product of the reaction?

(A) 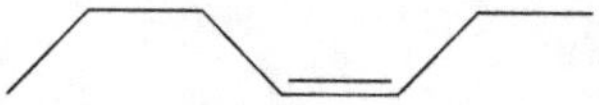 cis-alkene
(B) 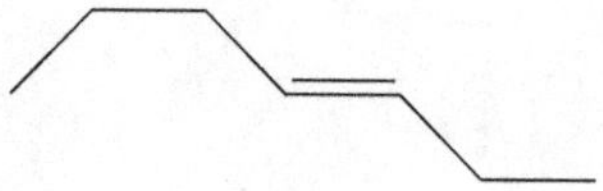 trans-alkene
(C) 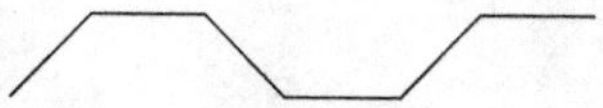 cis-alkene (alternate)
(D) 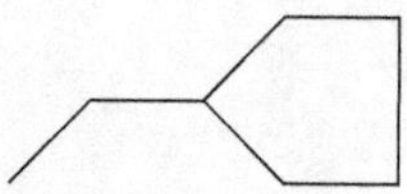 ethylcyclopentane

198. The following reaction produces one major product:

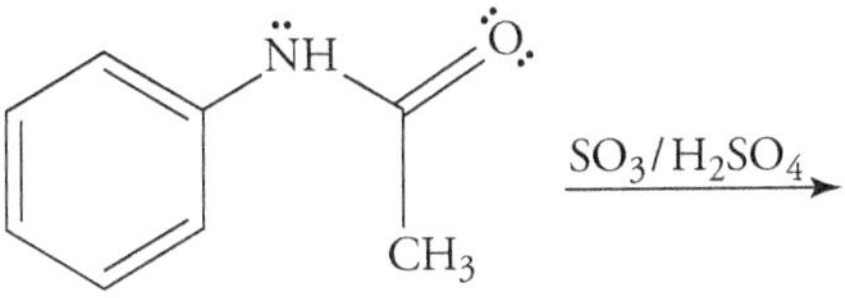

$$\text{C}_6\text{H}_5\text{-NH-CO-CH}_3 \xrightarrow{SO_3/H_2SO_4}$$

What is the major product of the reaction?

(A) *ortho*-substituted N-acetylaniline with $-S(=O)_2-OH$ group on nitrogen position shown on ring

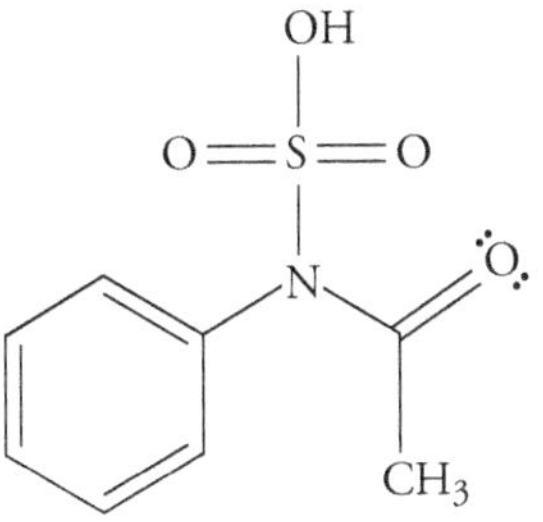

(B) N-acetylaniline with $-S(=O)_2-CH_2$-OH substituent

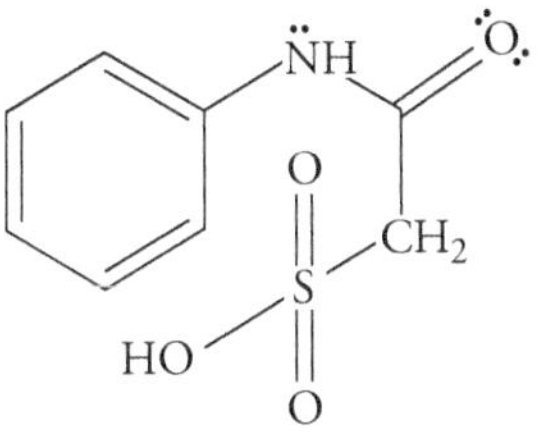

(C) *para*-substituted N-acetylaniline with $-SO_3H$ group (HO-S(=O)(=O)- on para position)

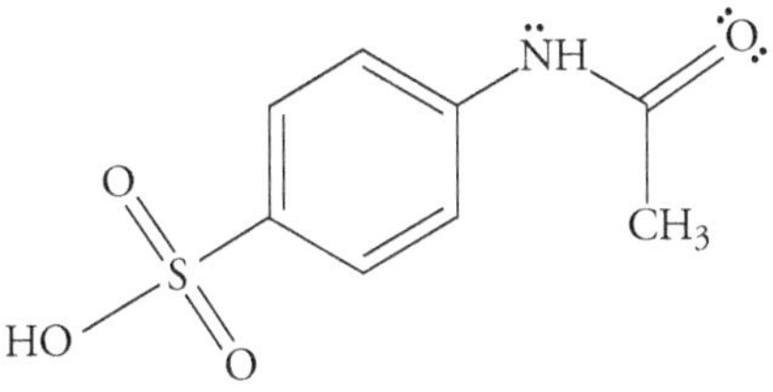

(D) *meta*-substituted N-acetylaniline with $-S(=O)_2-OH$ group

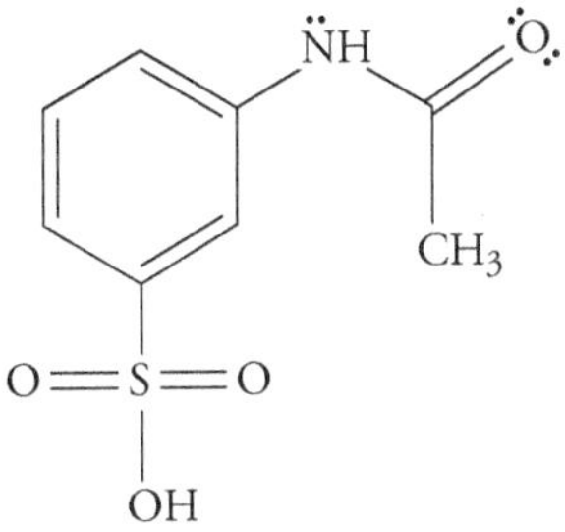

199. The following reaction produces one major product:

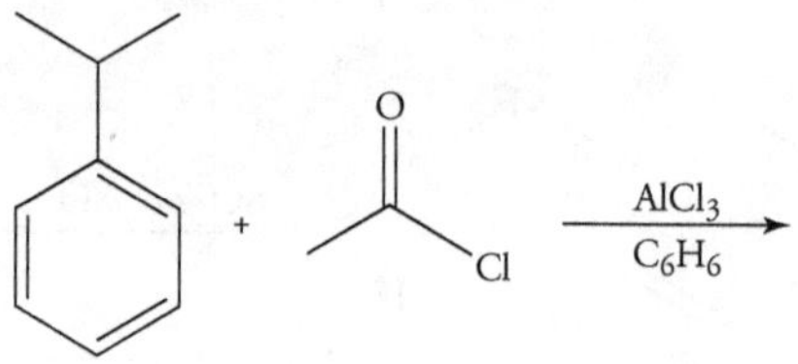

What is the major product of the reaction?

(A)

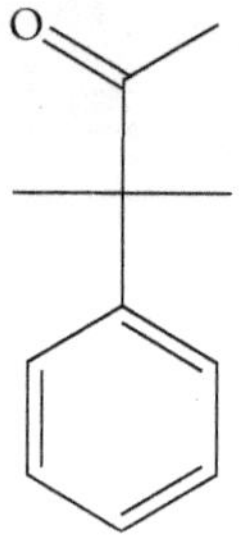

(B)

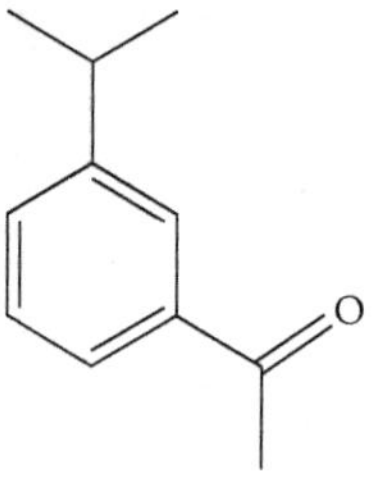

(C)

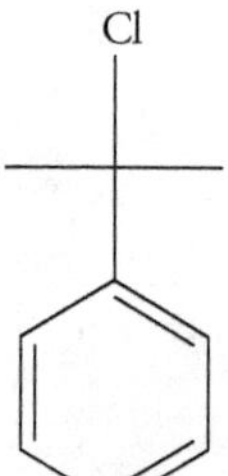

(D)

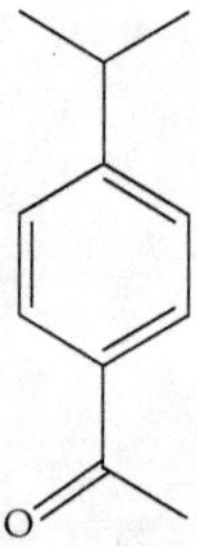

200. Assuming that all the following molecules are planar, which of the following is NOT aromatic?

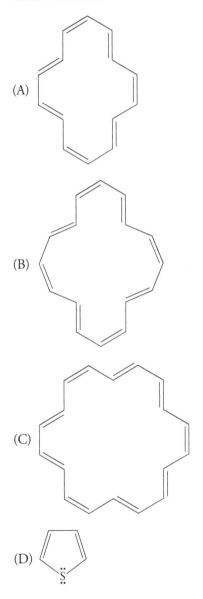

201. Which of the following species is NOT aromatic?

(A)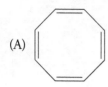

(B)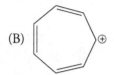

(C)

(D)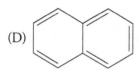

202. Which of the following is (are) the major product(s) formed by the reaction of aqueous bromine with 2-methylpropene?

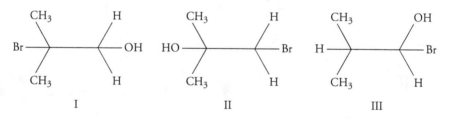

(A) I
(B) II
(C) III
(D) I and II

203. What product forms from the reaction of benzoic acid with a mixture of nitric and sulfuric acids?

(A) *para*-nitrobenzoic acid structure (COOH with NO₂ at para position)

(B) 2,4-dinitrobenzoic acid structure (COOH with NO₂ at ortho and para positions)

(C) *meta*-nitrobenzoic acid structure (COOH with NO₂ at meta position)

(D) 3-nitro-5-(sulfonyloxy)benzoic acid structure (COOH with O₂N and OSO₃ at 3,5 positions)

204. Where does substitution occur when nitrobenzene reacts with a hot mixture of Br$_2$ and FeBr$_3$?
    (A) Meta
    (B) Ortho
    (C) Para
    (D) Ortho and para

205. The addition of hydrogen bromide in the presence of peroxide to 1-pentene could form either of the products shown below.

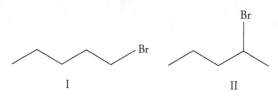

    I II

    How do the relative amounts of the two products compare?
    (A) Product I will predominate.
    (B) Product II will predominate.
    (C) There will be equal amounts of Products I and II.
    (D) Neither product will form.

206. The hydrogenation of the following compound using a Lindlar catalyst will form what type of compound?

    (A) An alkane
    (B) An alkene
    (C) An alkyne
    (D) A cyclic compound

207. Which of the following are formed by the reaction of meta-chloroperoxybenzoic acid (mCPBA) with 4-methylcyclohexene, followed by treatment with base?

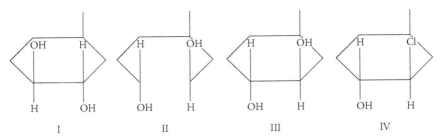

(A) I and III
(B) II and III
(C) III and IV
(D) I and II

208. Which of the following are formed by the reaction of basic permanganate with cyclohexene?

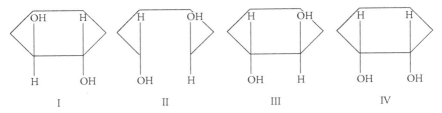

(A) II only
(B) I and III
(C) IV only
(D) II and IV

209. The addition of hydrogen bromide to 1-butene produces 2-bromobutane. The two products isolated are:

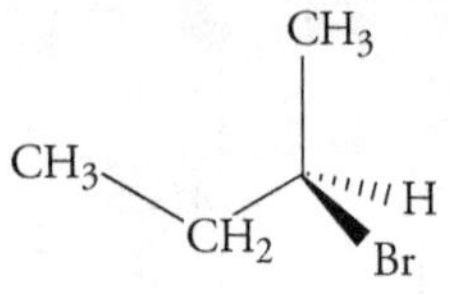

[Structure I: CH3-CH2-C(CH3)(H)(Br) with H on wedge back, Br on wedge front] [Structure II: mirror image with H on wedge back, Br on wedge front, CH2-CH3 on right]

I 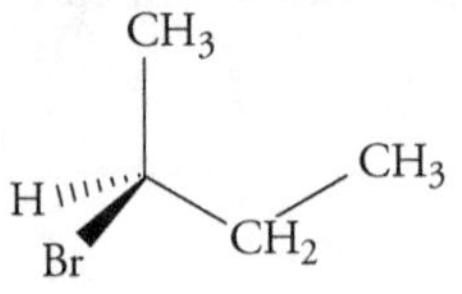   II

What is the best description of the products?
(A) Conformers
(B) Diastereomers
(C) Enantiomers
(D) Structural isomers

210. The addition of hydrogen bromide to 1-butene produces 2-bromobutane. The two products isolated in equal amounts are:

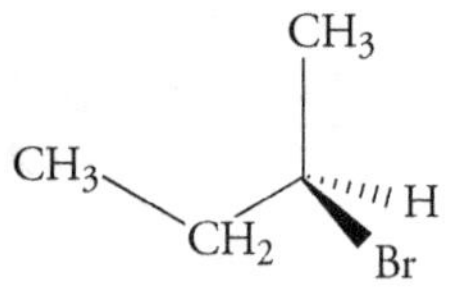

[Structure I] [Structure II]

I 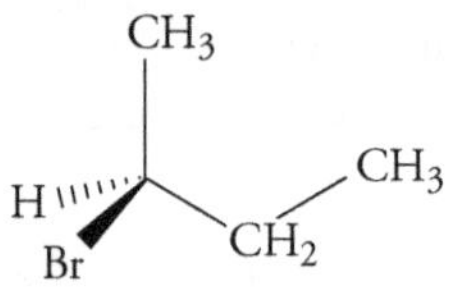   II

What is the specific rotation of the product mixture?
(A) 0°
(B) +15°
(C) −15°
(D) Impossible to determine

211. Which of the following best describes the addition of HBr to an alkene?
(A) It is a one-step process.
(B) It will form the anti-Markovnikov product.
(C) The process is initiated by nucleophilic attack.
(D) It is a two-step process.

212. The low reactivity of aromatic systems toward electrophilic substitution occurs because the aromatic system is:
(A) A poor nucleophile.
(B) A poor electrophile.
(C) Nonpolar.
(D) Very reactive.

213. The hydration of an alkene is an acid-catalyzed reaction that yields the Markovnikov alcohol. Why is the acid considered a catalyst?
    (A) The acid reacts in a 1:1 ratio with the alkene.
    (B) The acid increases the activation energy.
    (C) The acid is unnecessary, as the reaction will proceed without the acid.
    (D) The acid decreases the heat of reaction.

214. The exhaustive nitration of toluene yields 2,4,6-trinitrotoluene, better known as TNT. The formula for TNT is $C_7H_5N_3O_6$. What are the hybridizations of the nitrogen atoms?
    (A) Only sp^3
    (B) Some sp^2 and some sp^3
    (C) Only sp^2
    (D) Some sp and some sp^2

215. Which of the following reagents would convert the following compound to 2-bromo-2-methylbutane?

$$\underset{\underset{H}{|}}{\overset{\overset{CH_3}{|}}{C}}=\underset{\underset{CH_3}{|}}{\overset{\overset{CH_3}{|}}{C}}$$

    (A) Br_2
    (B) HBr
    (C) $Br_2 + H_2O$
    (D) HBr + peroxide

216. The mononitration of toluene yields a mixture of 2-nitrotoluene and 4-nitrotoluene. The formula of both compounds is $C_7H_7NO_2$. Both of these products may undergo further nitration to form dinitro and trinitro products. Further nitration requires a different temperature from that of the mononitration. Why is it necessary to change the temperature?
    (A) A higher temperature is needed because the first nitro group is ring-deactivating.
    (B) A higher temperature is needed because the first nitro group is ring-activating.
    (C) A lower temperature is needed because the first nitro group is ring-deactivating.
    (D) A lower temperature is needed because the first nitro group is ring-activating.

217. An organic chemist wishes to produce chloroheptane by the reaction of hydrogen chloride with cis-3-heptene. What would be the best solvent for this reaction?
    (A) cis-3-heptene
    (B) EtOH
    (C) CHOOH
    (D) H_2O

218. Which of the following dienophiles would react most rapidly with cyclopentadiene via a Diels-Alder reaction?

(A) 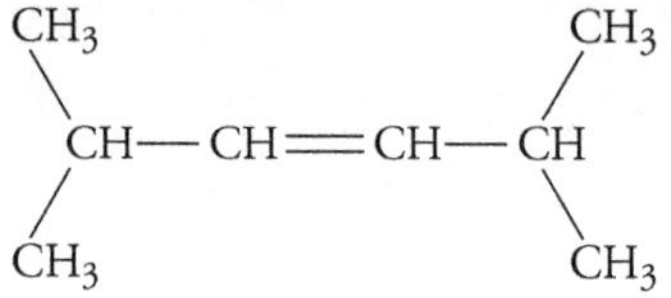
$$\begin{array}{c} CH_3 \\ | \\ CH-CH=CH-CH \\ | \\ CH_3 \end{array} \begin{array}{c} CH_3 \\ | \\ \\ | \\ CH_3 \end{array}$$

(B) $CH_3-CH=CH-O-CH_3$

(C) $CH_3-CH=CH-CH_3$

(D) 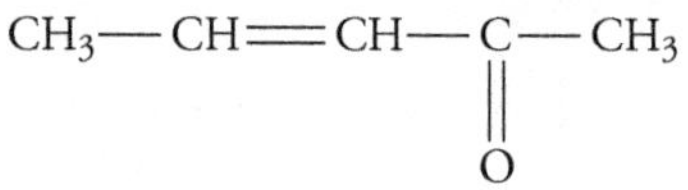 $CH_3-CH=CH-\underset{\underset{O}{\|}}{C}-CH_3$

219. An organic chemist conducts the following reaction:

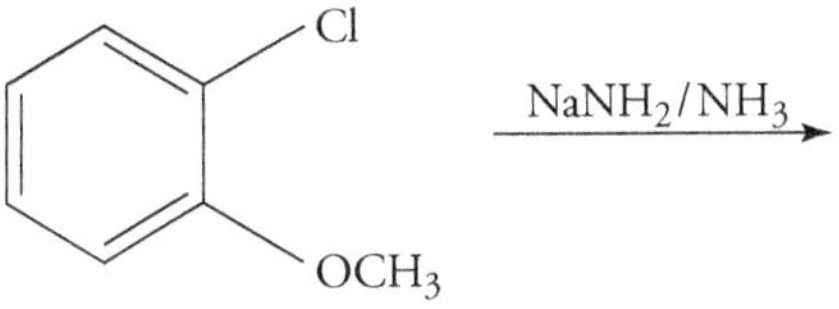

[Structure: benzene ring with Cl at position 1 and OCH₃ at position 2] $\xrightarrow{\text{NaNH}_2/\text{NH}_3}$

What is the major product?

(A) [Structure: benzene ring with Cl and NH₂ at adjacent positions]

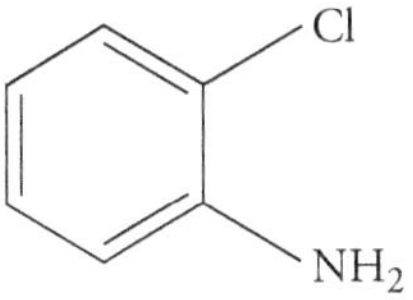

(B) [Structure: benzene ring with Cl, OCH₃, and H₂N substituents]

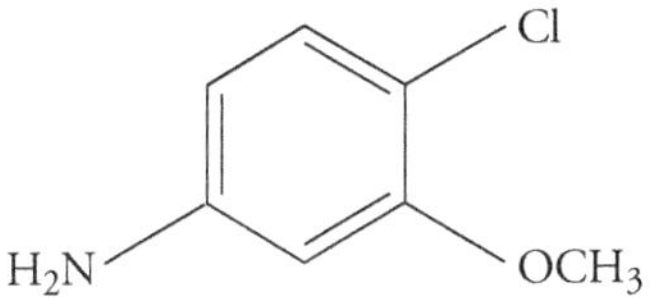

(C) [Structure: benzene ring with H₂N and OCH₃ in meta position]

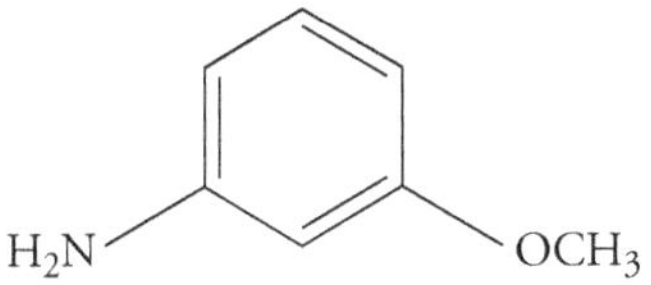

(D) [Structure: benzene ring with NH₂ and OCH₃ at adjacent positions]

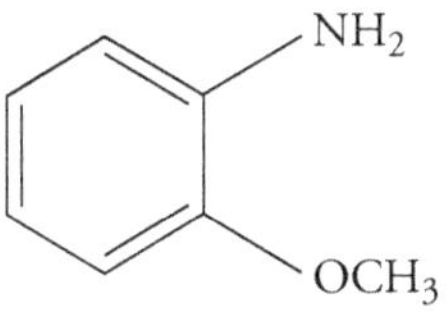

220. Which of the following compounds form from the reaction of 1-methylcyclohexene with B_2H_6 and H_2O_2/OH^-?

(A)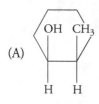

(B)

(C)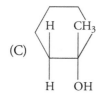

(D)

221. An organic chemist conducts four experiments on $CH_2=CH-C(CH_3)_3$. Which of the following reactions will lead to a change in the carbon backbone?
    (A) H_2SO_4/H_2O
    (B) B_2H_6/heat
    (C) Pt/H_2
    (D) HBr/benzoyl peroxide

222. Which of the following compounds would undergo sulfonation to produce the highest para-to-ortho sulfonic acid ratio?
    (A) Toluene
    (B) Isopropylbenzene
    (C) t-Butylbenzene
    (D) Ethylbenzene

223. An organic chemist treats the following compound with ozone and then treats the products with a zinc/water mixture. How many different products form?

(A) 4
(B) 3
(C) 2
(D) 1

224. Which of the following undergoes chlorination the fastest in the presence of $Cl_2/FeCl_3$?

(A) [benzene]

(B) [benzaldehyde]

(C) [benzoic acid]

(D) [methoxybenzene]

225. Electron-withdrawing groups deactivate aromatic rings toward reaction via electrophilic aromatic substitution. However, these same groups promote nucleophilic aromatic substitution. Which of the following is the best explanation for this observation?

(A) Resonance
(B) Steric hindrance
(C) Stabilization of the free radical
(D) Stabilization of the carbocation

226. Grignard reagents, such as $CH_3CH_2CH_2MgBr$, will react with carbonyl groups. Which of the following will react with a Grignard reagent to produce a secondary alcohol?

(A)
$$\underset{H}{\overset{O}{\underset{\|}{C}}}\underset{H}{}$$

(B)
$$CH_3CH_2\underset{}{\overset{O}{\underset{\|}{C}}}OH$$

(C)
$$CH_3CH_2\underset{}{\overset{O}{\underset{\|}{C}}}H$$

(D)
$$CH_3CH_2\underset{}{\overset{O}{\underset{\|}{C}}}CH_2CH_3$$

227. For most addition reactions of an alkene with a hydrogen halide, HX, the rate law is: rate = k[HX] [alkene]. This rate law indicates all of the following EXCEPT:

(A) The reaction is first-order with respect to the alkene.
(B) The reaction is second-order overall.
(C) The rate-determining step involves one molecule of HX and one molecule of the alkene.
(D) The reaction occurs in a single step involving one molecule of HX and one molecule of the alkene.

228. When an aromatic compound undergoes electrophilic aromatic substitution, the aromatic substrate is the:
    (A) Nucleophile.
    (B) Electrophile.
    (C) Dienophile.
    (D) Intermediate.

229. The mononitration of toluene yields a mixture of 2-nitrotoluene and 4-nitrotoluene. The formula of both compounds is $C_7H_7NO_2$. The two products are examples of:
    (A) Conformers.
    (B) Enantiomers.
    (C) Constitutional isomers.
    (D) Geometric isomers.

230. Alkenes will undergo an acid-catalyzed addition of water. Which of the following is NOT involved?
    (A) There is the formation of a carbocation intermediate.
    (B) The anti-Markovnikov product will form.
    (C) The reaction results in the formation of an alcohol.
    (D) The initial step is the protonation of the double bond.

231. The addition of hydrogen bromide to 1-butene produces 2-bromobutane. The two products isolated in equal amounts are:

    Which of the following is NOT true about the reaction?
    (A) The result of the reaction is a racemic mixture.
    (B) The rate-determining step in the mechanism is not the formation of a carbocation.
    (C) The overall rate law is second-order.
    (D) It is possible to separate the products by fractional distillation.

232. One reaction of alkenes is ozonolysis. The reaction sequence begins with the reaction with ozone and ends with a treatment of the ozonide formed with zinc and water. How many different products will the ozonolysis of the following compound yield?

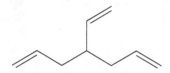

(A) 2
(B) 5
(C) 3
(D) 4

233. The addition of hydrogen bromide to 1-pentene could form either of the following products:

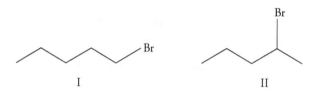

How do the relative amounts of the two products compare?
(A) I only.
(B) II only.
(C) Equal amounts of each.
(D) The distribution depends upon the temperature.

234. The addition of bromine in carbon tetrachloride to 1-pentene forms 1,2-dibromopentane:
(A) As a meso compound.
(B) As a syn addition product.
(C) As an enantiomer of the reactant.
(D) As a racemic mixture.

235. The following reaction produces one major product:

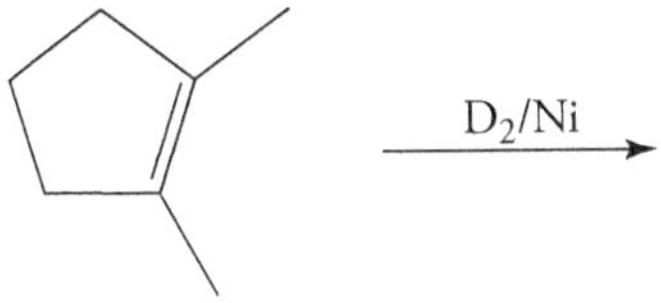

$$\text{1,2-dimethylcyclopentene} \xrightarrow{D_2/Ni}$$

What is the major product of the reaction?

(A) 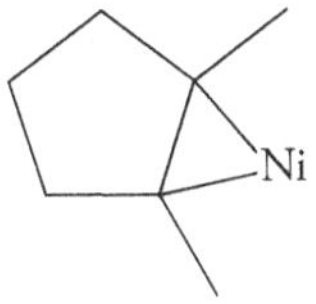 [structure with Ni bridge]

(B) 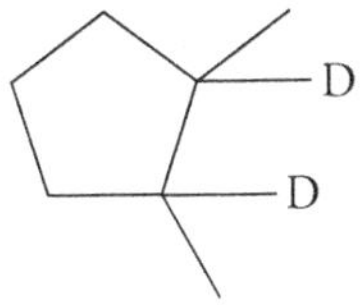 [structure with both D on wedge bonds, same face]

(C) 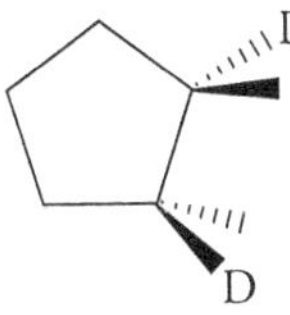 [structure with one D on wedge, one D on dash — trans]

(D) [structure with both D on dash bonds, same face]

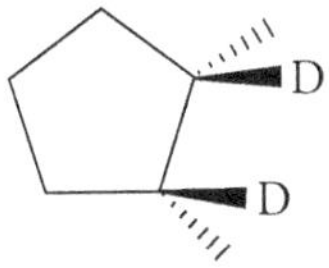

236. The following reaction produces one major product:

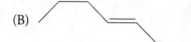
1) Na/NH₃(l)
2) H₂O

What is the major product of the reaction?

(A)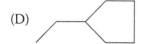

(B)

(C)

(D)

237. The following reaction produces one major product:

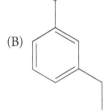

What is the major product of the reaction?

(A)

(B)

(C)

(D)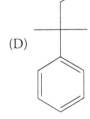

238. An organic chemist is studying the following Diels-Alder reaction:

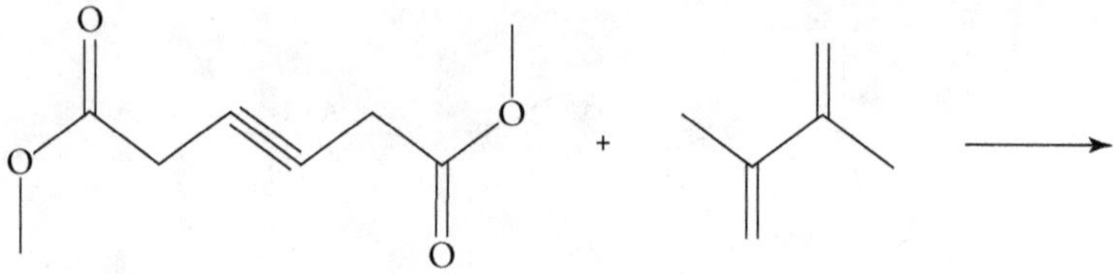

What is the major product?

(A)

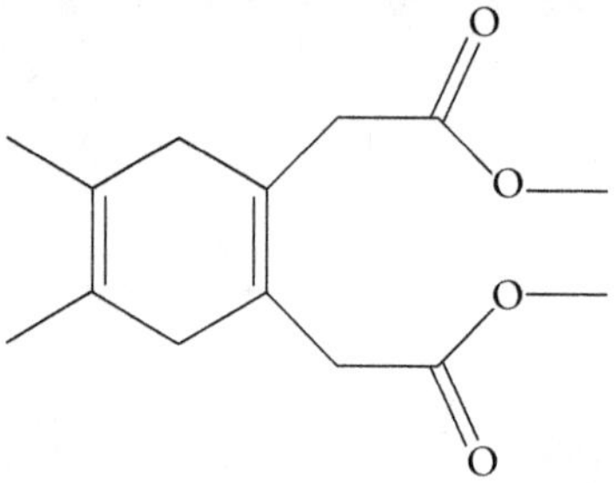

(B)

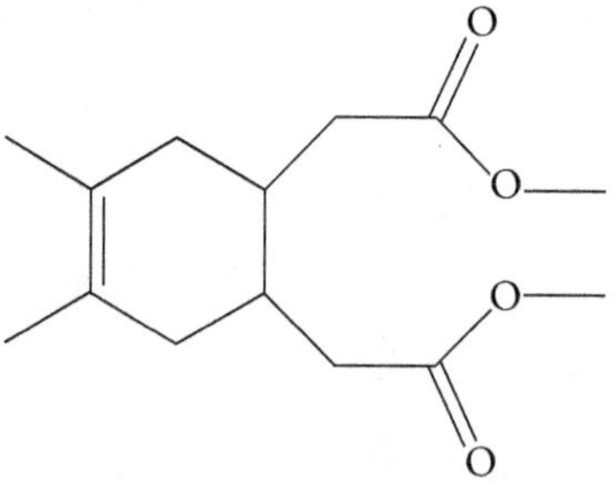

(C)

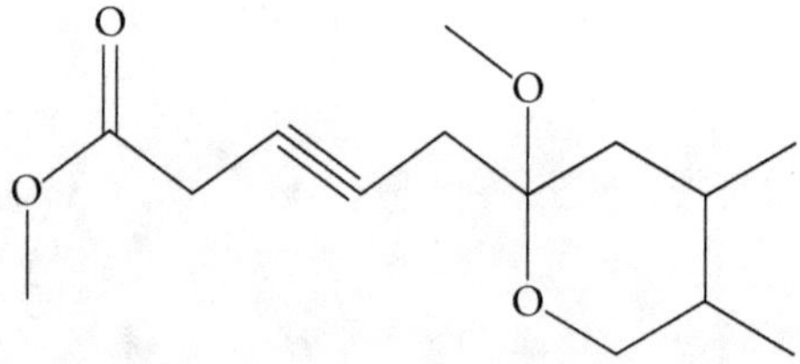

(D)

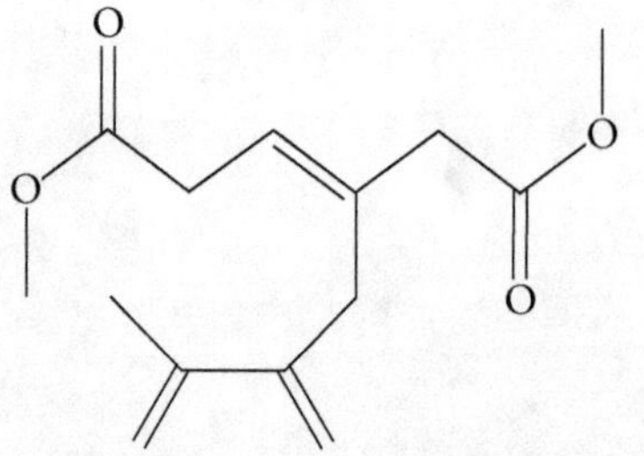

239. Sodium ethoxide will react with 3-chloro-3-ethylpentane at room temperature. An organic chemist does two experiments, each with equal amounts of the two reactants. In one experiment, 0.050 mole of each reactant is used, and the reaction takes 10 minutes to go to completion. In the second experiment, the amount of sodium ethoxide is doubled. How long will it take the second reaction to go to completion?

(A) 5 minutes
(B) 10 minutes
(C) 20 minutes
(D) 7.5 minutes

240. The hydration of an alkene may lead to two different compounds. What is the relationship between these two compounds?

(A) They are tautomers.
(B) They are conformers.
(C) They are regioisomers.
(D) They are anomers.

241. An organic chemistry student investigates the following reaction in lab:

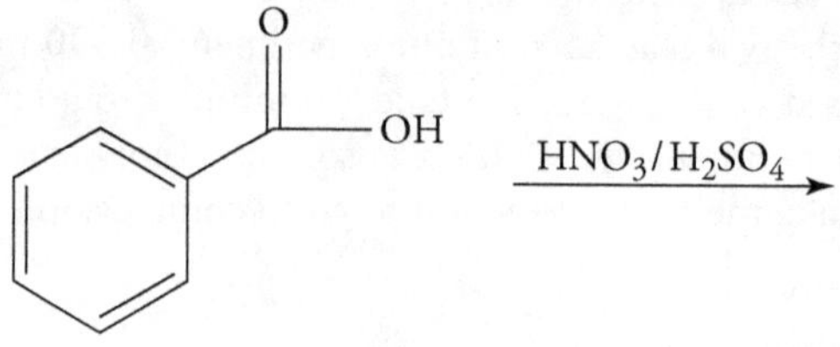

C₆H₅–C(=O)–OH → HNO₃/H₂SO₄

Which of the following is the major product?

(A) nitrobenzene (NO₂ on benzene ring)

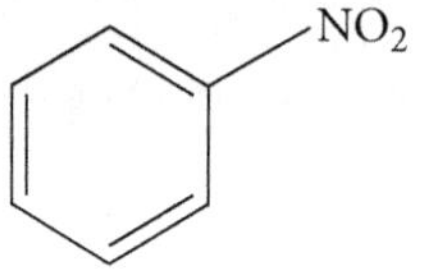

(B) 2-nitrobenzoic acid (ortho -NO₂ to -COOH)

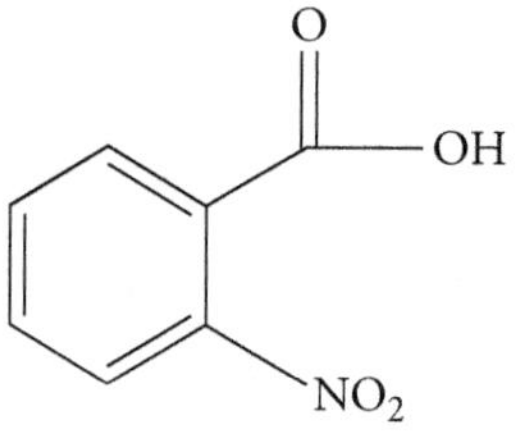

(C) 4-nitrobenzoic acid (para -NO₂ to -COOH)

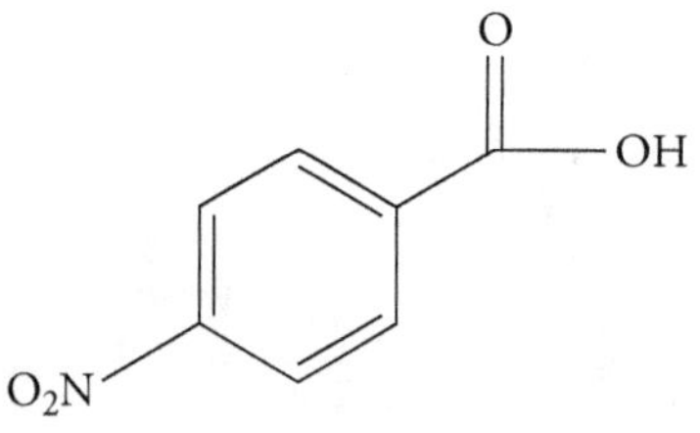

(D) 3-nitrobenzoic acid (meta -NO₂ to -COOH)

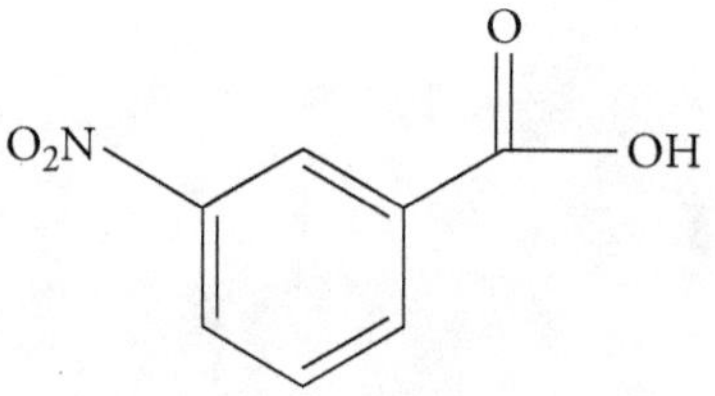

242. An organic chemistry student performs the following experiment in lab:

$$\text{C}_6\text{H}_5\text{–NO}_2 \xrightarrow{\text{Zn(Hg)}, \text{HCl}} \text{C}_6\text{H}_5\text{–NH}_2$$

What is the purpose of the zinc amalgam, Zn(Hg)?
(A) It is a reducing agent.
(B) It is an oxidizing agent.
(C) It is a catalyst.
(D) It is a Lewis base.

CHAPTER 5

Nucleophilic and Cyclo Addition Reactions

Questions 243–246 refer to the following passage.

The reaction of a carbonyl (aldehyde or ketone) with a base will produce an enolate ion (a nucleophile). This nucleophile will attack any electrophile. If we add a base to a carbonyl with no electrophile present, the reaction will still occur because the carbonyl group itself is an electrophile; as the enolate forms, it can attack the carbonyl group of another aldehyde or ketone molecule. This is called an aldol reaction. In the aldol reaction, aldehydes or ketones reacting with α-hydrogen atoms in the presence of dilute base give an aldol (a β-hydroxyaldehyde or ketone). An aldol contains an aldehyde (or ketone) and an alcohol group in the β-position. The aldol formed in this reaction is 2-methyl-3-hydroxypentanal.

The aldol formed by the aldol reaction, especially if heated, can react further. The heating will cause dehydration (loss of H_2O). The overall reaction involving an aldol reaction followed by dehydration is the aldol condensation. The product of an aldol condensation is an α,β-unsaturated aldehyde (an enal) or ketone. The presence of extended conjugation favors the formation of this product. This process works better if extended conjugation results. The aldol reaction and condensation are reversible.

243. The following is an example of an aldol addition:

$$CH_3-\underset{\underset{H}{\|}}{C}=O \quad \xrightarrow{10\% \text{ NaOH}/H_2O}_{5°C} \quad \underset{CH_3}{\overset{OH}{\underset{|}{CH}}}-\underset{CH_2}{CH}-\underset{H}{\overset{O}{\underset{\|}{C}}}$$

What is the role of the NaOH in this reaction?

(A) Deprotonation
(B) Protonation
(C) Electrophile
(D) Termination

244. What is the first step in an aldol condensation under basic conditions?
    (A) Removal of the α-hydrogen
    (B) Protonation of a carbonyl group
    (C) Reaction of the enol
    (D) Addition of base to the carbonyl

245. The following is part of the aldol condensation involving an enolate ion and an aldehyde:

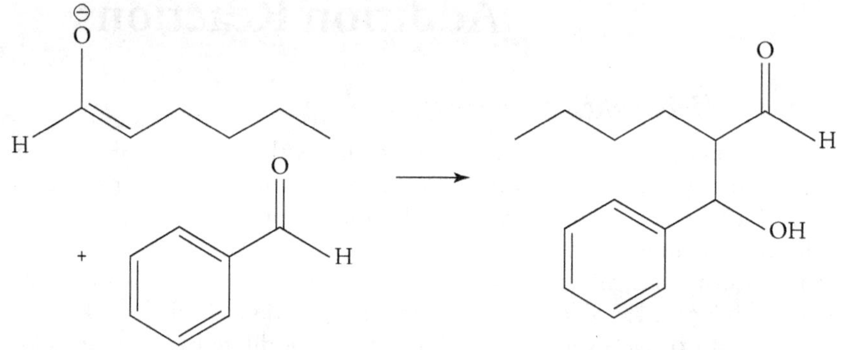

    What is the role of the enolate ion?
    (A) It is a nucleophile.
    (B) It is an electrophile.
    (C) It is a catalyst.
    (D) It is a free radical.

246. What is the final product of the following reaction?

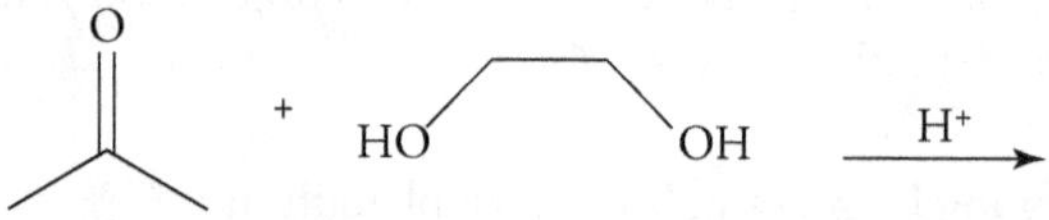

    (A) An ether
    (B) A hemiacetal
    (C) An acetal
    (D) A carboxylic acid

247. What is the name of the process in which an aldehyde or ketone is in equilibrium with its enol form?
    (A) Mutarotation
    (B) Inversion of configuration
    (C) Tautomerism
    (D) Hydrolysis

248. While investigating the following chemical reaction, an organic chemist found that one product formed in a 95% yield.

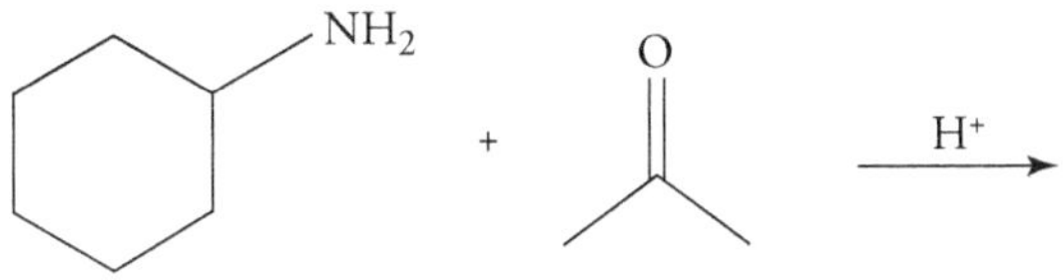

cyclohexyl-NH₂ + acetone —H⁺→

What was the structure of the one product formed?

(A) 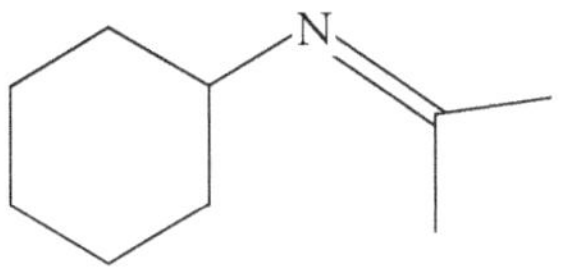 cyclohexyl-N=C(CH₃)₂

(B) 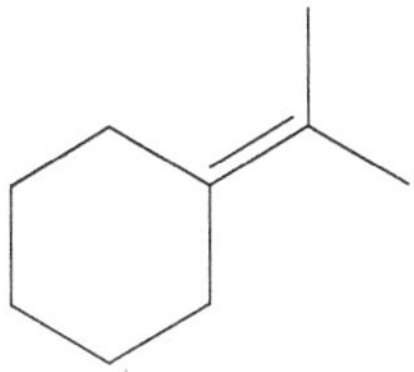 cyclohexylidene=C(CH₃)₂

(C) cyclohexyl-O-CH(CH₃)₂

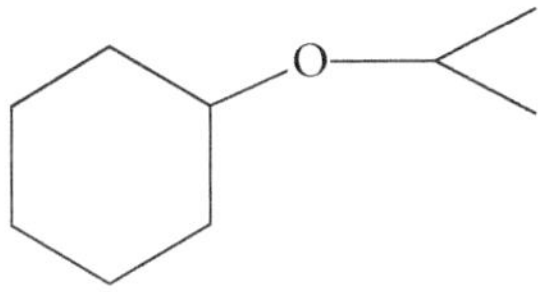

(D) 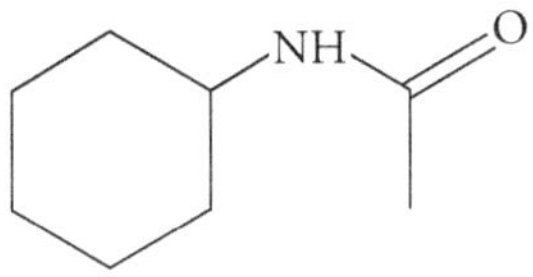 cyclohexyl-NH-C(=O)CH₃

249. While investigating the following chemical reaction, an organic chemist found that one product formed in a 75% yield.

Br–CH₂–CH₂–CH₂–CH₂–CH(OH)–CH₂–OH + (CH₃)₂C=O →[H⁺, −H₂O]

What was the structure of the one product formed?

(A) [structure: branched acetal with O-C(CH₃)₂-O bridging two oxygens on a chain with Br]

(B) [structure: Br-chain with 1,3-dioxolane ring containing C(CH₃)₂]

(C) Br–CH₂–CH₂–CH₂–CH₂–CH=CH₂

(D) [structure: cyclohexane with Br, O-CH(CH₃)₂, and OH substituents]

250. While investigating the following chemical reaction, an organic chemist found that one product formed in a 95% yield.

What was the structure of the one product formed?

(A) [structure: pentyl-O-C(CH3)2]

(B) [structure: butyl-Mg-O-C(CH3)2]

(C) [structure: pentyl with OH and methyl branch]

(D) [structure: with Br and Mg-O-C(CH3)2]

251. An organic chemist was investigating the following reaction:

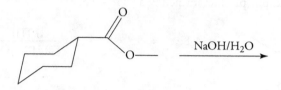

What was the structure of one of the organic products?

(A)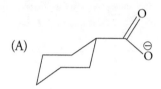

(B)

(C)

(D)

252. An organic chemist was investigating the following reaction:

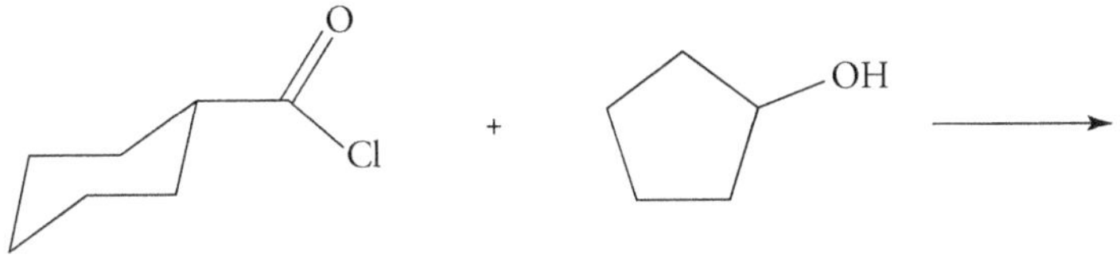

[Cyclohexyl-C(=O)-Cl] + [cyclopentyl-OH] ⟶

What was the structure of the major product?

(A) Cyclohexyl-C(=O)-cyclopentyl

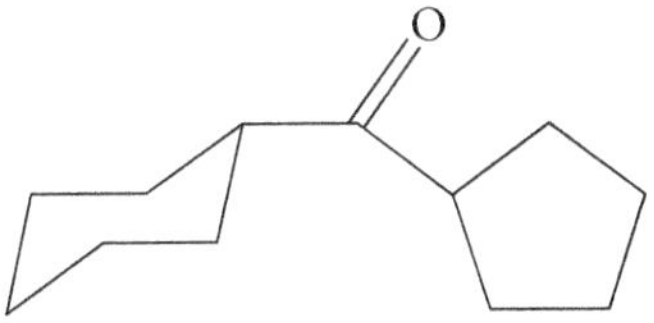

(B) Cyclohexyl-C(=O)-O-cyclopentyl

(C) Cyclohexyl-C(=O)-C(Cl)(cyclopentyl)

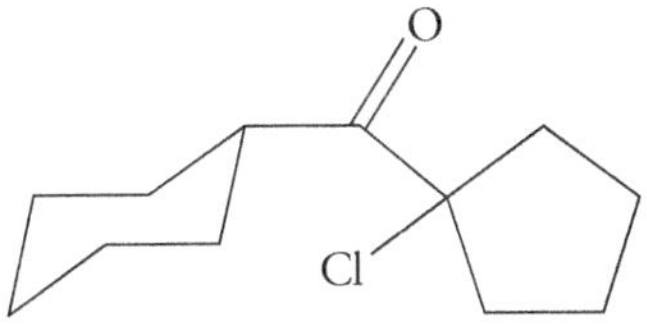

(D) Cyclohexyl-C(=O)-O-C(Cl)(cyclopentyl)

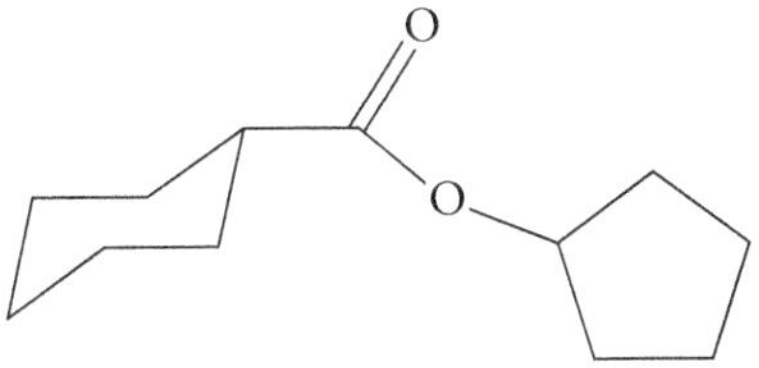

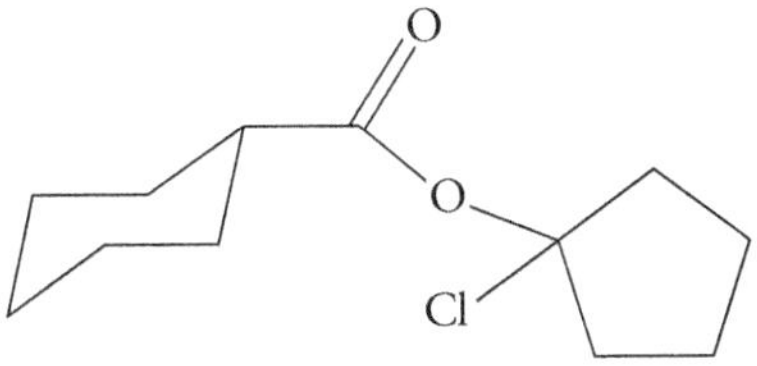

253. It is possible to generate a carboxylic acid through the acid hydrolysis of an ester. Which of the following esters would produce 3-methylpentanoic acid upon acid hydrolysis?

(A), (B), (C), (D)

254. It is possible to oxidize many alcohols to carboxylic acids by oxidation with acidic potassium dichromate. This method of synthesis will work for which of the following alcohols?

(A) benzyl alcohol structure with OH

(B) 3-pentanol structure with OH

(C) 2-methyl-2-pentanol structure with OH

(D) phenol structure with OH

255. It is possible to reduce 3-methylbutanoic acid with LiAlH$_4$. The product of the LiAlH$_4$ reaction will react with additional 3-methylbutanoic acid, in the presence of an acid catalyst, to produce a new compound. What is the structure of the new compound?

(A)

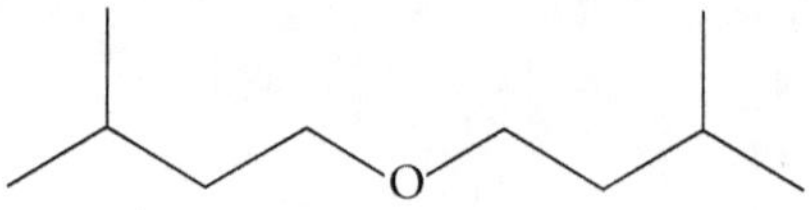

(B)

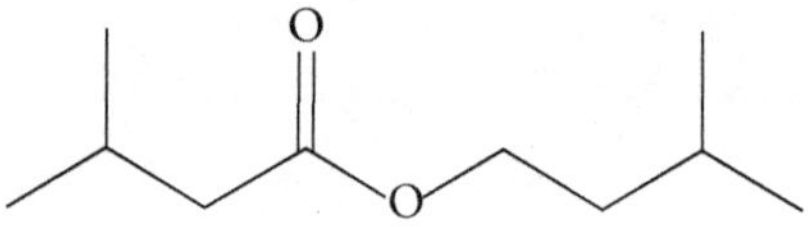

(C)

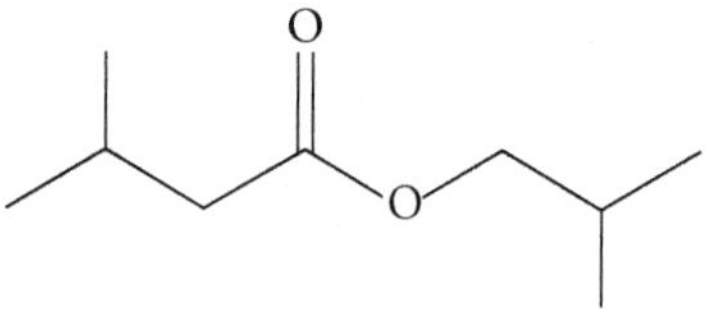

(D)

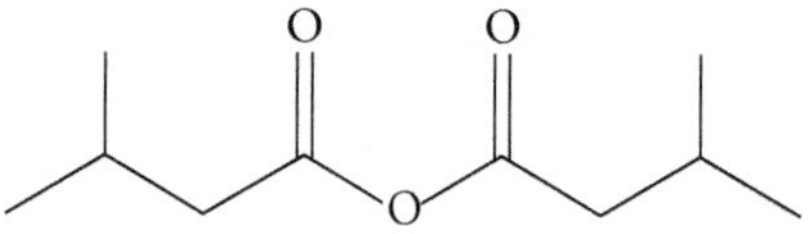

256. Which of the following is the strongest acid?

(A) 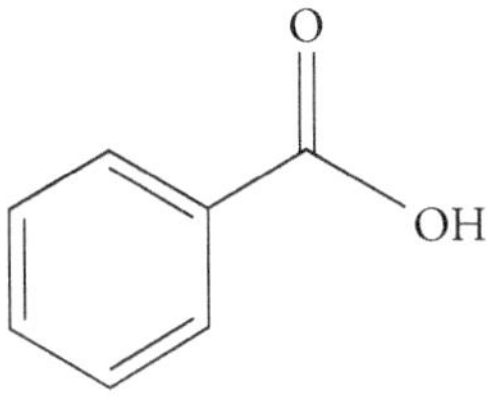 benzoic acid

(B) 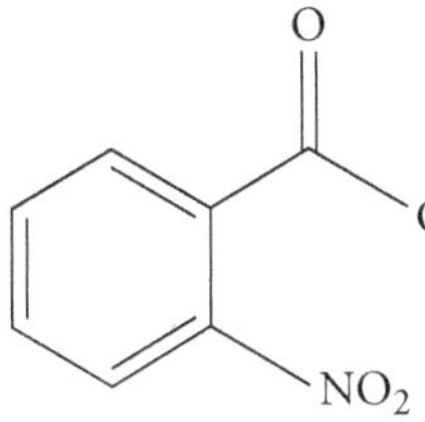 2-nitrobenzoic acid

(C) 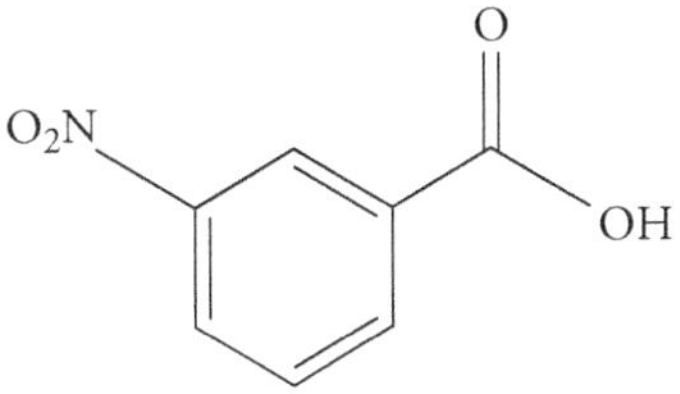 3-nitrobenzoic acid

(D) 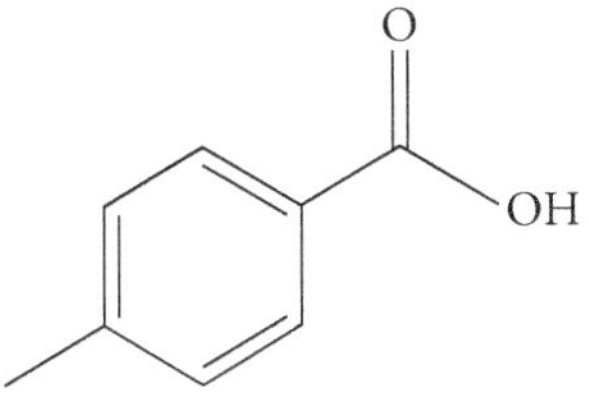 4-methylbenzoic acid

257. In organic synthesis, it is possible to replace acid chlorides with acid bromides. What is the advantage of using the more expensive acid bromide?
  (A) It is easier to separate the products.
  (B) The large bromide ion minimizes side reactions.
  (C) The bromide released catalyzes further reaction.
  (D) Acid bromides are more reactive.

258. Which of the following compounds contains a hemiacetal?

(A) [tetrahydrofuran ring with -OH substituent on carbon adjacent to ring O]

(B) [tetrahydrofuran ring with -O-CH₃ substituent on carbon adjacent to ring O]

(C) [CH₃-C(=O)-O-CH₃ structure]

(D) [1,3-dioxolane ring]

259. What is the best explanation why acyl halides react by addition-elimination reactions while aldehydes and ketones undergo nucleophilic addition reactions?

(A) Acyl halides are stronger electrophiles.
(B) The halide ions are good leaving groups.
(C) Acyl halides are more polar.
(D) The carbonyl group is more reactive if there is no halide attached.

260. Which of the following is a characteristic of hemiacetal formation through the reaction of an alcohol with an aldehyde?

(A) The transition state is 5-coordinate.
(B) It has a carbocation intermediate.
(C) It is a stereospecific reaction.
(D) It follows second-order kinetics.

261. What is the major product of the reaction of excess propanal with aqueous base at 0–10°C?

(A) [structure: CH3CH2-C(=O)-CH(CH3)-C(=O)-H]

(B) [structure: CH3CH2-CH(OH)-CH(CH3)-C(=O)-H]

(C) [structure: 1,4-cyclohexanedione]

(D) [structure: CH3CH2-CH(OH)-CH2-C(=O)-CH3]

262. What type of reaction occurs when the enolate anion of one carbonyl compound reacts with a different carbonyl compound?
 (A) An aldol condensation
 (B) An esterification
 (C) A saponification
 (D) A neutralization

263. Rank the following acids in order of increasing acidity.

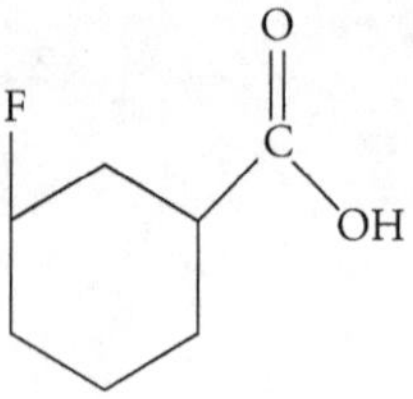

[Structures: I - cyclohexane with F on ring and CH(COOH); II - cyclohexane with CH(COOH); III - cyclohexane with CH(COOH) and F on ring; IV - cyclohexane with C(COOH)(F)]

  I 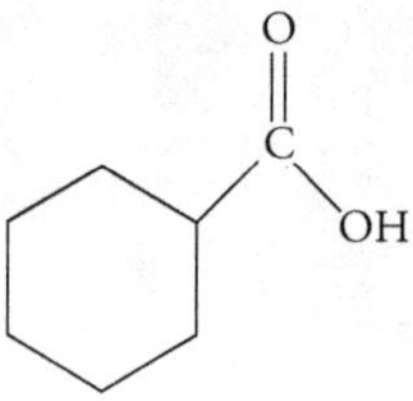　　　　　　　　II 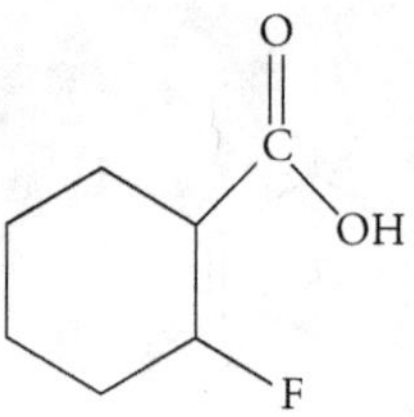　　　　　　　　III 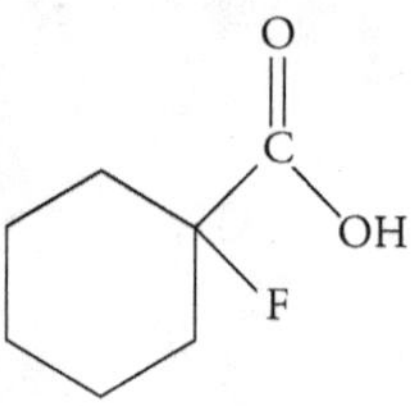　　　　　　　　IV

(A) III < I < IV < II
(B) IV < III < I < II
(C) I = III < II < IV
(D) II < I < III < IV

264. Which of the following combinations, in the presence of acid, will result in the formation of an acetal?

(A) Ketone plus an aldehyde
(B) Carboxylic acid plus an alcohol
(C) Carboxylic acid plus an acetate ion
(D) Ketone plus an alcohol

265. Which of the following compounds is the strongest acid?

(A)  [1-butanol structure] OH

(B) 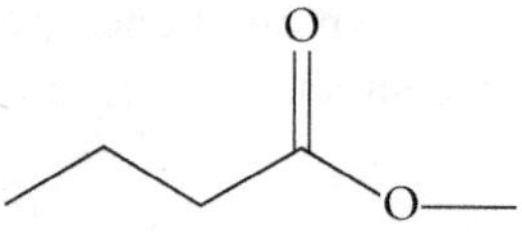 [butanoate ester structure]

(C) 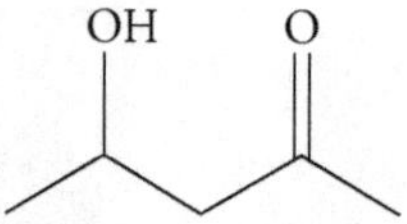 [4-hydroxy-2-butanone type structure with OH and O]

(D) 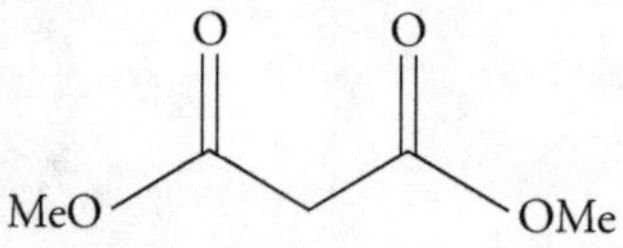 [dimethyl malonate: MeO-C(=O)-CH2-C(=O)-OMe]

266. What happens when an aldehyde and a dilute acid are mixed?
    (A) A hydroxy aldehyde forms through an aldol condensation.
    (B) A carboxylic acid forms through oxidation.
    (C) A ketone forms through a nucleophilic attack followed by oxidation.
    (D) A carboxylic acid forms through tautomerism.

267. The acid-catalyzed hydrolysis of an amide will produce which of the following?
    (A) An imide and an aldehyde
    (B) An amine and a carboxylic acid
    (C) An ammonium ion and a carboxylic acid
    (D) An amine and a carboxylate ion

268. What type of compound is the major product of the following reaction?

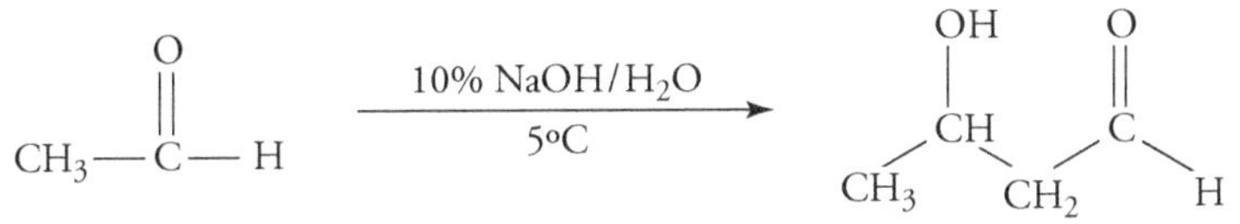

$$CH_3-\overset{O}{\underset{\|}{C}}-H \xrightarrow[5°C]{10\% \text{ NaOH/H}_2\text{O}} CH_3-\underset{\underset{OH}{|}}{CH}-CH_2-\overset{O}{\underset{\|}{C}}-H$$

    (A) A diketone
    (B) A ketone
    (C) An aldehyde
    (D) An ester

269. Which of the following compounds will NOT react with a very strong base such as potassium t-butoxide?

(A) 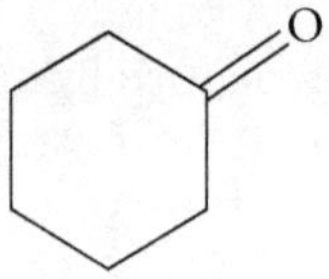 [cyclohexanone]

(B) 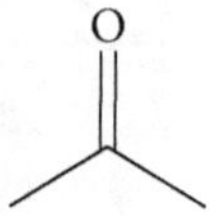 [acetone]

(C) 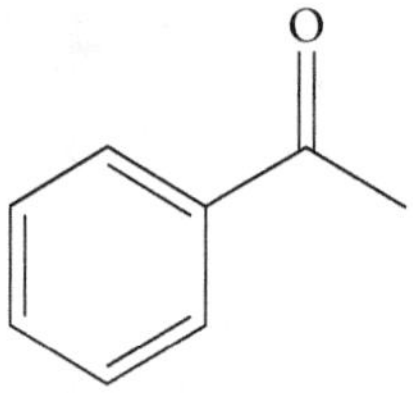 [acetophenone]

(D) 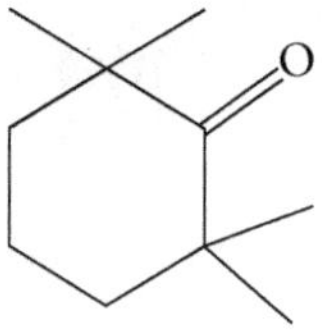 [2,2,6-trimethylcyclohexanone-type structure]

270. What is the best procedure for the following reaction?

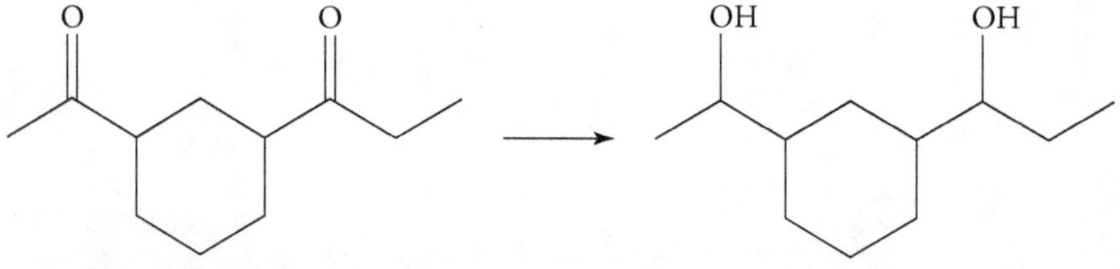

[diketone on cyclohexane → diol on cyclohexane]

(A) NaBH$_4$/MeOH 25°
(B) ZnCl$_2$/HCl
(C) 1) LiAlH$_4$ 2) HCl
(D) 1) B$_2$H$_6$ 2) H$_2$O$_2$/OH$^-$

271. Which of the following acids is the easiest to decarboxylate?

(A) 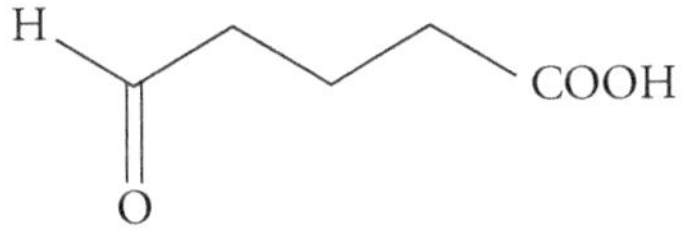 OHC-CH₂-CH₂-CH₂-COOH

(B) 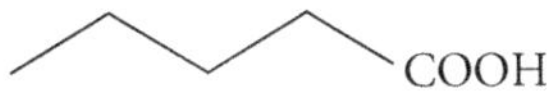 CH₃-CH₂-CH₂-COOH

(C) 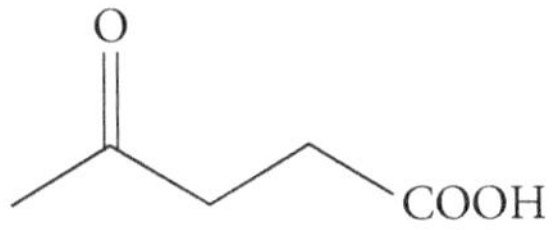 CH₃-CO-CH₂-CH₂-COOH

(D) 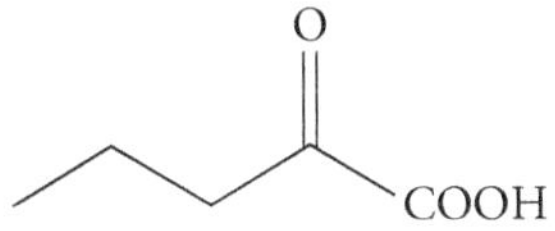 CH₃-CH₂-CO-CH₂-COOH

272. What is the major product of the reaction of excess ethylmagnesium bromide with ethyl butanoate followed by acid workup?

(A) 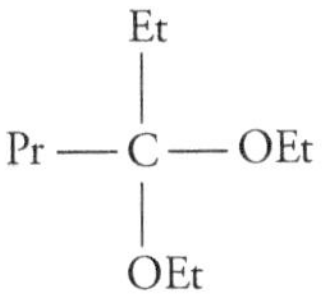 Pr—C(Et)(OEt)—OEt

(B) Pr—C(=O)—Et

(C) Pr—C(Et)(OH)—Et

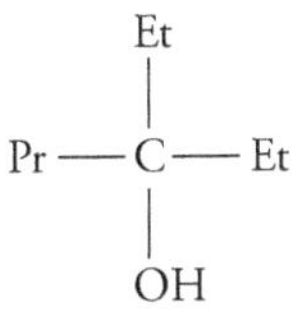

(D) 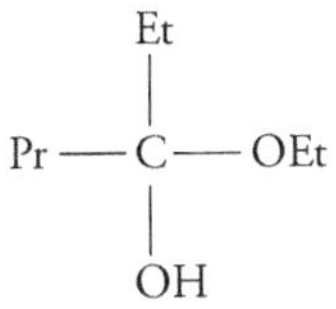 Pr—C(Et)(OEt)—OH

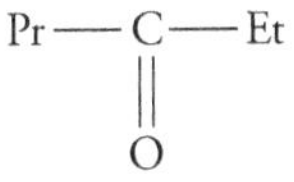

273. Which of the following is part of the mechanism for the following reaction?

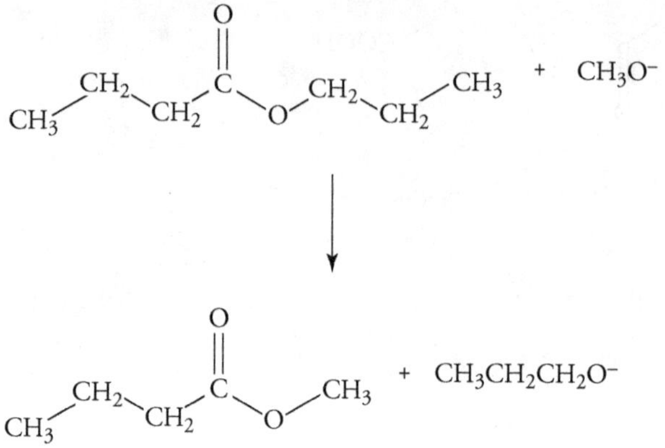

$$CH_3-CH_2-CH_2-C(=O)-O-CH_2-CH_2-CH_3 + CH_3O^-$$

↓

$$CH_3-CH_2-CH_2-C(=O)-O-CH_3 + CH_3CH_2CH_2O^-$$

(A) Attack by CH_3O^-, then elimination of $CH_3CH_2CH_2O^-$
(B) Elimination of $CH_3CH_2CH_2O^-$, then attack by CH_3O^-
(C) One-step backside attack by CH_3O^-
(D) Through a carbonium ion intermediate

274. The following compound reacts with a hydrochloric acid solution:

$$\begin{array}{c} CH_3 \\ \diagdown CH-O \\ | \diagdown CH-CH_2-CH_2F \\ CH \diagup \\ CH_3\diagup O \end{array}$$

What are the products of the reaction?

(A)
$$\begin{array}{c} CH_3 \\ \diagdown CH-O \\ | \diagdown CH-CH_2-CH_2Cl \\ CH \diagup \\ CH_3\diagup O \end{array}$$

(B)
$$\begin{array}{c} CH_3 \\ \diagdown CH-OH \\ | \\ CH \\ CH_3\diagup \diagdown OH \end{array} \quad \begin{array}{c} O \\ \| \\ HC-CH_2-CH_2F \end{array}$$

(C)
$$\begin{array}{c} CH_3 \\ \diagdown C=O \\ \diagup \\ C=O \\ CH_3\diagup \end{array} \quad \begin{array}{c} O \\ \| \\ C-CH_2-CH_2F \\ HO\diagup \end{array}$$

(D)
$$\begin{array}{c} CH_3 \\ \diagdown C=O \\ \diagup \\ C=O \\ CH_3\diagup \end{array} \quad \begin{array}{c} O \\ \| \\ HC-CH_2-CH_2F \end{array}$$

275. Which of the following compounds will most easily react with $H_2N(CH_2)_4NH_2$ to form a polymer?

(A) $Cl-\underset{\underset{O}{\|}}{C}-CH_2-\underset{\underset{O}{\|}}{C}-Cl$

(B) $Cl-\underset{\underset{O}{\|}}{C}-CH_2-\underset{\underset{O}{\|}}{C}-H$

(C) $CH_3-\underset{\underset{O}{\|}}{C}-CH_2-\underset{\underset{O}{\|}}{C}-Cl$

(D) $Cl-\underset{\underset{O}{\|}}{C}-CH_2-\underset{\underset{O}{\|}}{C}-O-CH_3$

Nucleophilic and Cyclo Addition Reactions 129

276. An organic chemist produces the following compound by ozonolysis after an acid workup:

$$CH_3-\underset{\underset{O}{\parallel}}{C}-CH_2-\underset{CH_3}{CH}-CH_2-\underset{\underset{O}{\parallel}}{C}-CH_3$$

Which of the following compounds was the starting material?

(A) $CH_3-\underset{OH}{CH}-CH_2-\underset{CH_3}{CH}-CH_2-\underset{OH}{CH}-CH_3$

(B)

(cyclopentene with CH₃ and H on one sp² carbon, CH–CH₃ substituent, CH and CH₂ ring members, CH₃ substituent)

(C) $CH_3-\underset{\underset{O}{\parallel}}{C}-CH_2-\underset{CH_2}{\overset{\parallel}{C}}-CH_2-\underset{\underset{O}{\parallel}}{C}-CH_3$

(D)

(cyclic structure with CH₃, C=C, CH₂, CH–CH₃, CH₂, C, CH₃)

277. Which of the following compounds will NOT react with ethylamine to produce the following compound?

(benzamide with N-ethyl substituent: C₆H₅–C(=O)–NH–CH₂CH₃)

(A) Benzaldehyde
(B) Benzoyl chloride
(C) Benzoic anhydride
(D) Benzoic acid

278. What are the products of the reaction of isopropyl phenyl ether with hot concentrated hydrobromic acid?

(A) Isopropyl bromide plus phenol
(B) Isopropyl alcohol and phenol
(C) Isopropyl alcohol plus bromobenzene
(D) Isopropyl bromide plus bromobenzene

279. An organic chemistry student mixes the following chemicals:

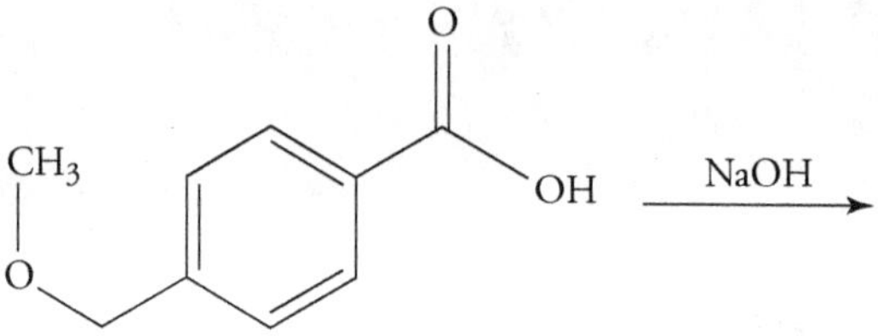

Which of the following is the product of this reaction?

(A)

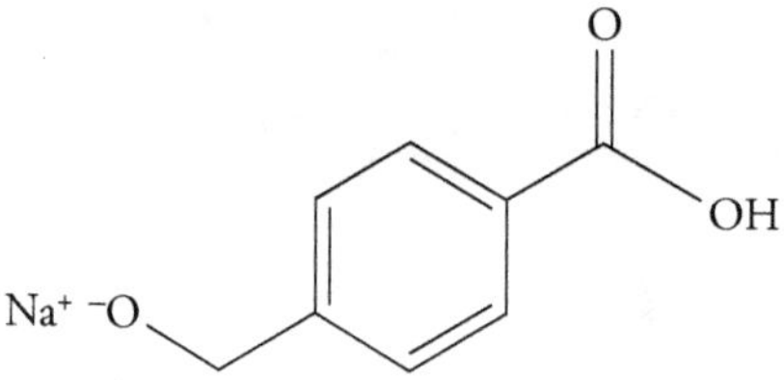

(B)

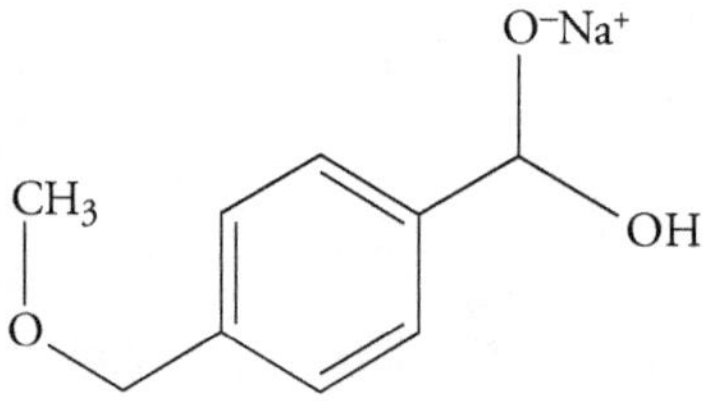

(C)

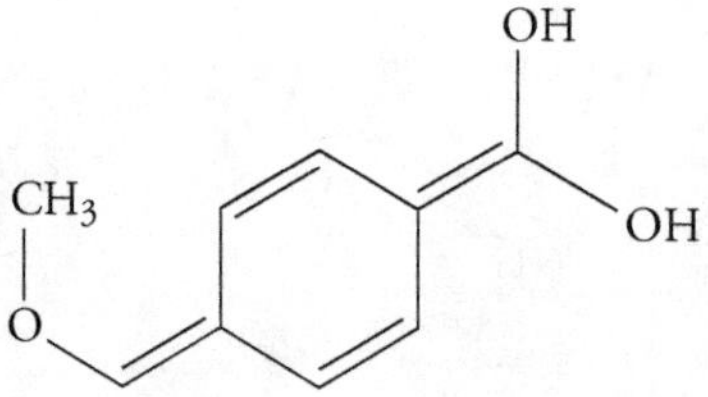

(D)

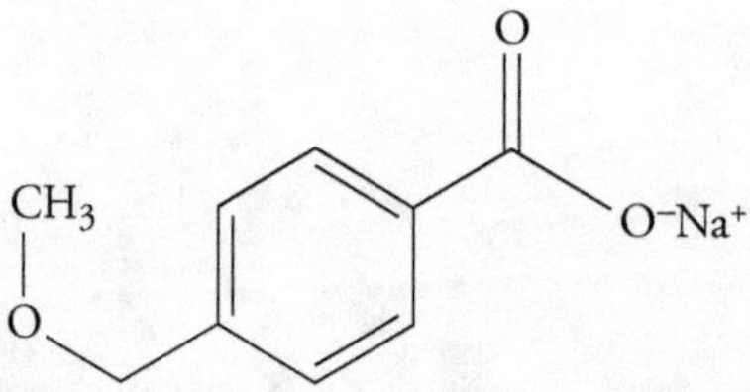

280. The following reaction occurs upon heating. This step is an example of what type of reaction?

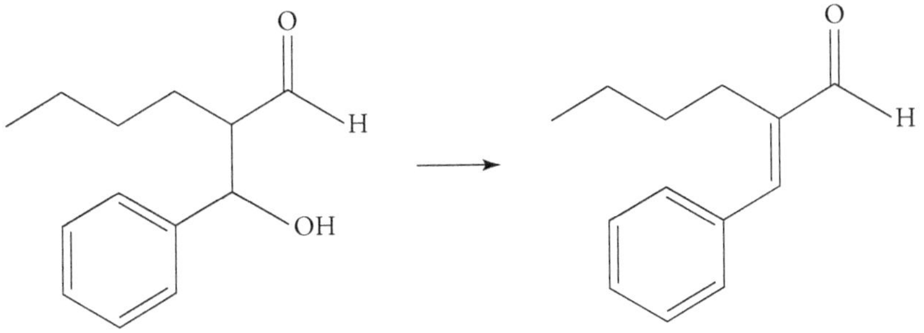

(A) Addition
(B) Elimination
(C) Substitution
(D) Dehydrogenation

281. An organic chemist is studying the following reaction:

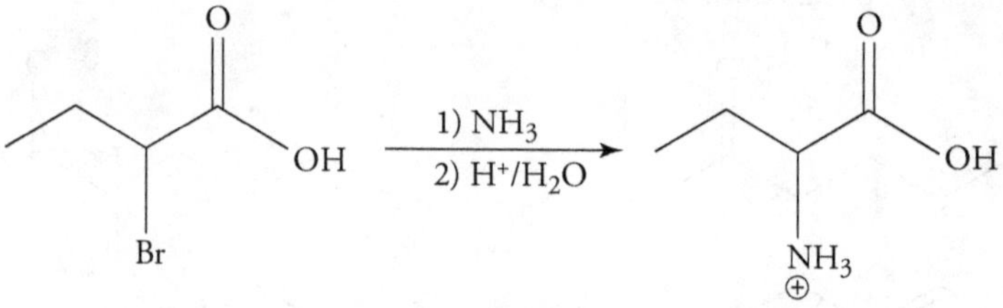

[structure: 2-bromobutanoic acid] + 1) NH$_3$ 2) H$^+$/H$_2$O → [structure: 2-aminobutanoic acid with NH$_3^+$]

During one experiment, he did not have sufficient NH$_3$. What would be the product of the first step if only one equivalent of NH$_3$ were available?

(A) 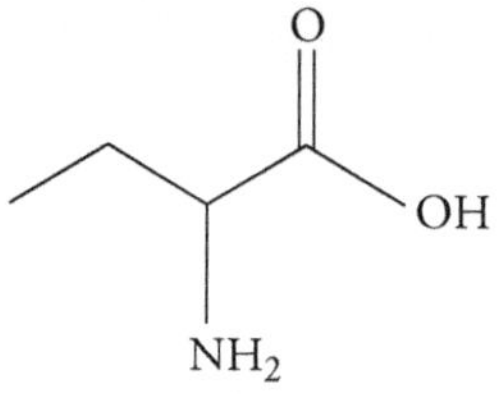 [structure with NH$_2$ and OH]

(B) 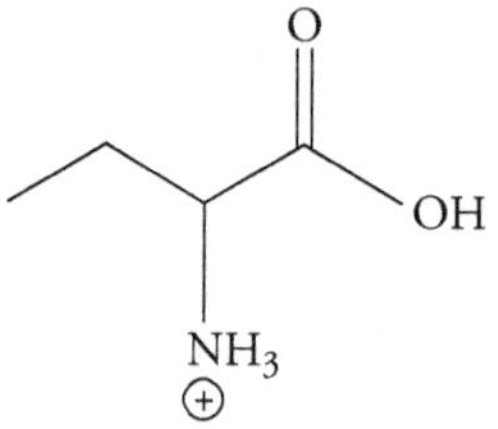 [structure with NH$_3^+$ and OH]

(C) 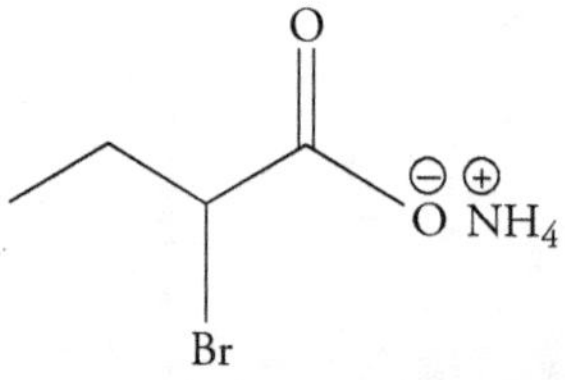 [structure with Br and O$^-$ NH$_4^+$]

(D) 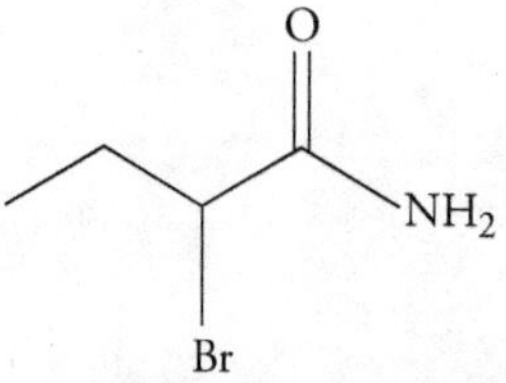 [structure with Br and NH$_2$ (amide)]

282. Which of the following reagents will serve to induce the following reaction?

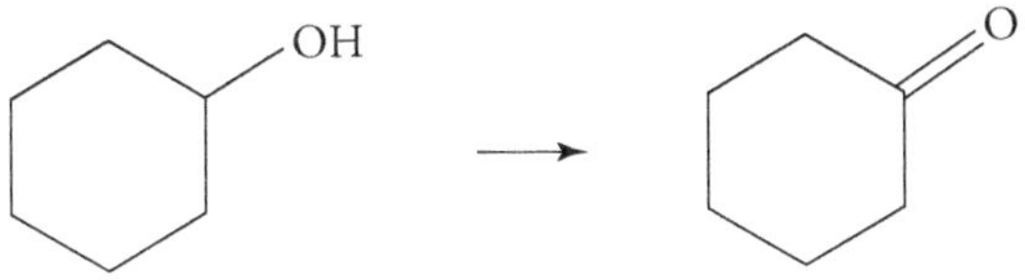

(A) CrO_3
(B) $LiAlH_4$
(C) PBr_3
(D) H_3PO_4

283. The Barbier reaction involves the reaction of an alkyl halide with an aldehyde or ketone in the presence of a variety of reactants, such as Mg, Li, or Zn. The generic reaction is:

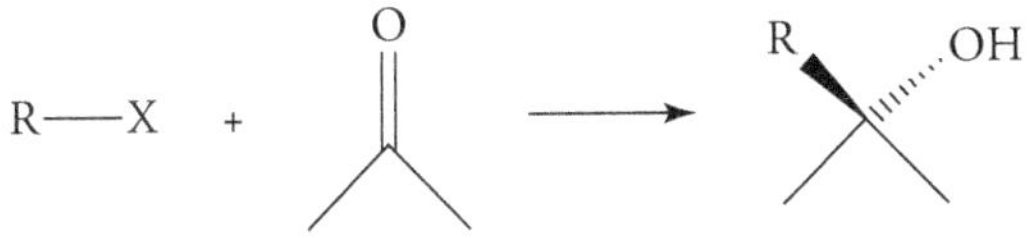

What type of reaction is this?
(A) Markovnikov addition
(B) Oxidation
(C) Nucleophilic substitution
(D) Reduction

284. The peptide bonds in proteins are stabilized by resonance. Which of the following is LEAST effective in stabilizing the peptide bond?

(A) H\N—C⫽O (with substituents)

(B) H\N—C with N having ⊖ and O having ⊕

(C) H\N⊕=C with O having ⊖

(D) H\N⊕=C with O having ⊖

285. The compound $C_{12}H_{14}O$ has a strong infrared (IR) absorption near 1700 cm^{-1}. The treatment of the compound with excess EtMgBr followed by a workup using aqueous acid gave a new compound. Where would the new compound show a strong absorption in its IR spectrum?
  (A) 3350 cm^{-1}
  (B) 1700 cm^{-1}
  (C) 3000 cm^{-1}
  (D) 1650 cm^{-1}

286. Compound I has the formula $C_{22}H_{24}O$, and it has a strong IR absorption near 1700 cm^{-1}. The treatment of the compound with excess EtMgBr followed by a workup using aqueous acid gave a new compound, II. Treatment of Compound II with acidic potassium dichromate gave Compound III. Compound III had the formula $C_{24}H_{24}O_2$. Which of the three compounds could be extracted by aqueous sodium bicarbonate?
  (A) I only
  (B) II and III only
  (C) III only
  (D) I and II only

287. Traditionally, soaps are prepared from fats. Soaps contain the carboxylate ion of a fatty acid. The best way to produce a soap from a fat is to heat the fat with:
    (A) Aqueous base.
    (B) Aqueous acid.
    (C) Chlorine.
    (D) Potassium permanganate.

288. What is the best reactant for producing butyl acetate from butyl alcohol?
    (A) CH_3COOH
    (B) CH_3CH_2OH
    (C) $CH_3COO^-Na^+$
    (D) $(CH_3CO)_2O$

289. What type of compound forms from the acid-catalyzed reaction of ethanol with benzoic acid?
    (A) An ester
    (B) Another acid
    (C) A ketal
    (D) An ether

290. What will be the product of the reaction of chloroacetic acid with one equivalent of ammonia?
    (A) $ClCH_2CONH_2$
    (B) H_2NCH_2COOH
    (C) $ClCH_2COO^-NH_4^+$
    (D) $H_2NCH_2COO^-$

291. The following is a simplified illustration of a Diels-Alder reaction:

During this reaction, there is a change in the number of σ bonds and π bonds. What is the overall change in each type of bond?
    (A) $+2\ \sigma$ and $-2\ \pi$
    (B) $-2\ \sigma$ and $+2\ \pi$
    (C) $+3\ \sigma$ and $-3\ \pi$
    (D) $-3\ \sigma$ and $+3\ \pi$

292. The saponification of an ester may produce the following products:

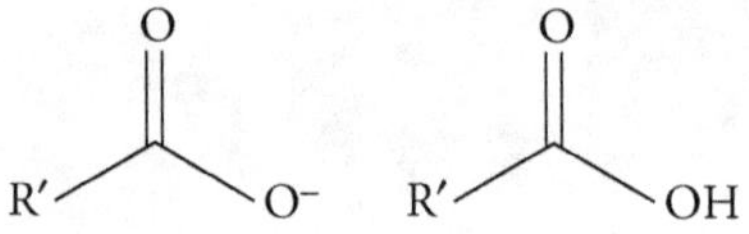

R'−C(=O)−O⁻ (I)   R'−C(=O)−OH (II)   ROH (III)   RO⁻ (IV)

Which are the most likely products to form?
(A) II and III
(B) I and III
(C) I and IV
(D) II and IV

293. Which of the hydrogen atoms in the following compound is the most acidic?

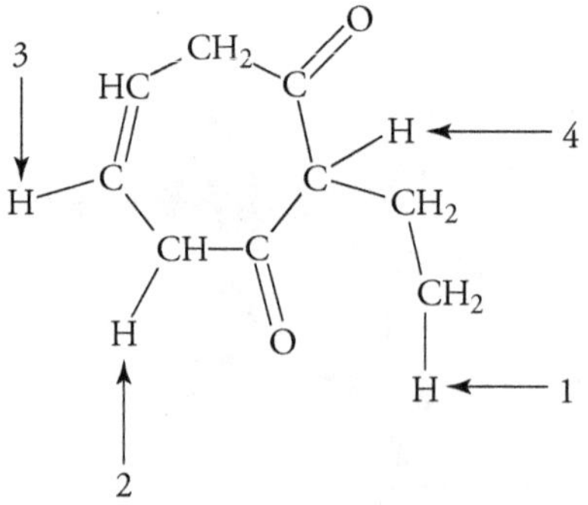

(A) 4
(B) 3
(C) 2
(D) 1

294. Rank the following nitrophenols in order of increasing acidity.

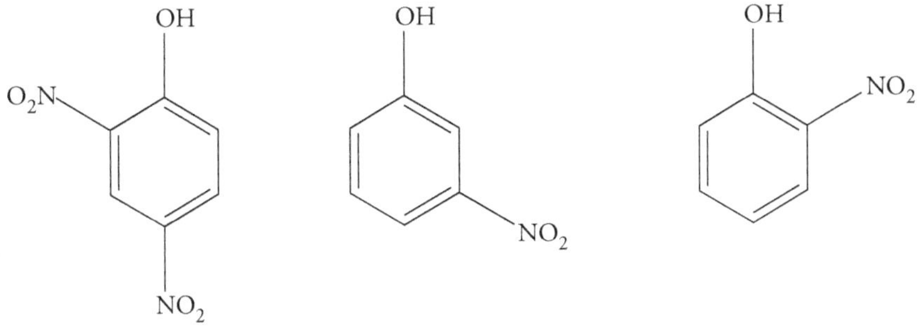

(A) II < I < III
(B) I < II < III
(C) III < II < I
(D) II < III < I

295. Classify each of the following compounds as an acetal, a hemiacetal, or neither.

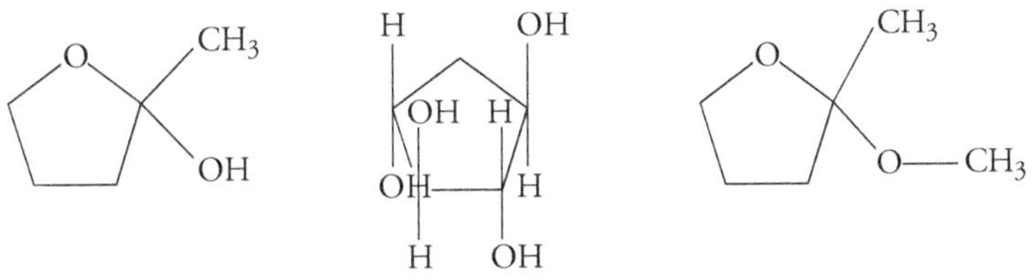

(A) I = neither, II = acetal, and III = hemiacetal.
(B) I = acetal, II = hemiacetal, and III = neither.
(C) I = hemiacetal, II = neither, and III = acetal.
(D) I = acetal, II = neither, and III = hemiacetal.

296. The experimental determination of the molar mass of acetic acid vapor results in a value of about 120 g/mol. What is the best explanation of this observation?
(A) Acetic acid undergoes an esterification reaction.
(B) Acetic acid undergoes a condensation reaction when it vaporizes.
(C) Acetic acid dehydrates to form acetic anhydride in the vapor phase.
(D) Acetic acid is present as hydrogen-bonded dimers.

297. The following nucleophile readily adds to the β-carbon because the π-electrons are delocalized. The increased electron density is localized on which atom?

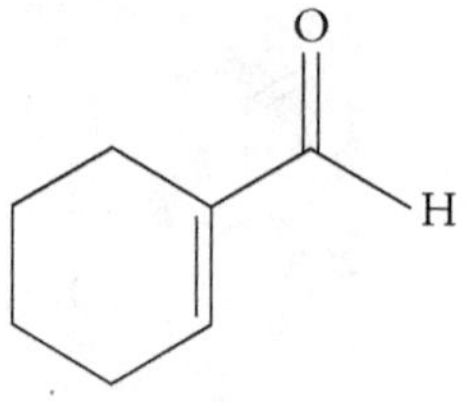

(A) The carbonyl oxygen
(B) The carbonyl carbon
(C) The γ-carbon
(D) The α-carbon

298. The following compound, benzoyl chloride, reacts with water to produce benzoic acid:

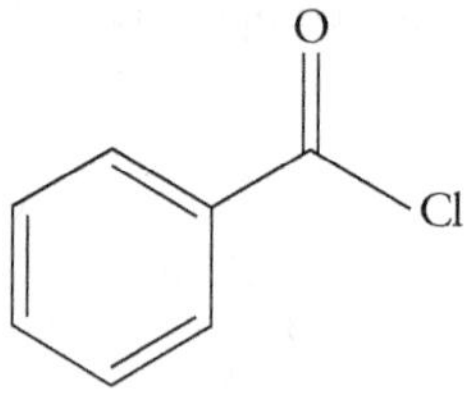

This is an addition-elimination reaction in which water:
(A) And benzoyl chloride are nucleophiles, and the chloride ion is the electrophile.
(B) Is the electrophile, benzoyl chloride is the nucleophile, and the chloride ion is the leaving group.
(C) And benzoyl chloride are electrophiles, and the chloride ion is the nucleophile.
(D) Is the nucleophile, benzoyl chloride is the electrophile, and the chloride ion is the leaving group.

299. What is the major product of the reaction of excess propanal with hot aqueous base?

(A) [structure: CH₃CH₂-C(=O)-CH(CH₃)-C(=O)-H]

(B) [structure: cyclohexane-1,4-dione]

(C) [structure: CH₃CH=C(CH₃)-C(=O)-H]

(D) [structure: CH₃CH₂-CH=CH-C(=O)-CH₃]

300. An organic chemist is investigating the following reaction:

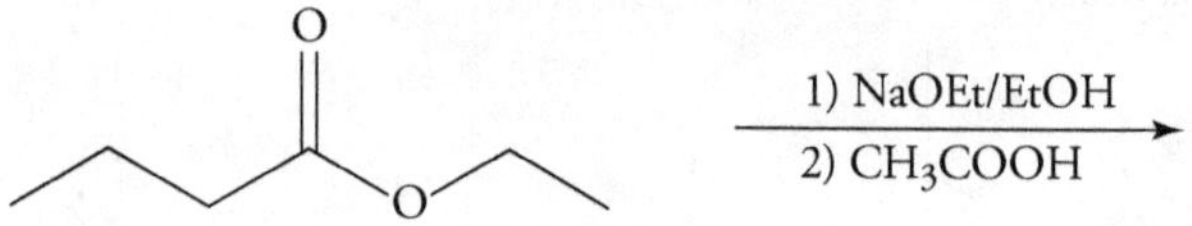

ethyl butanoate → 1) NaOEt/EtOH 2) CH₃COOH

What is the major product of the reaction?

(A) 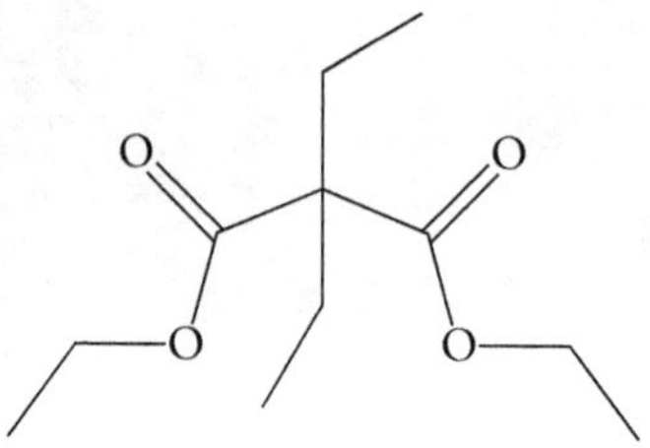 [structure]

(B) 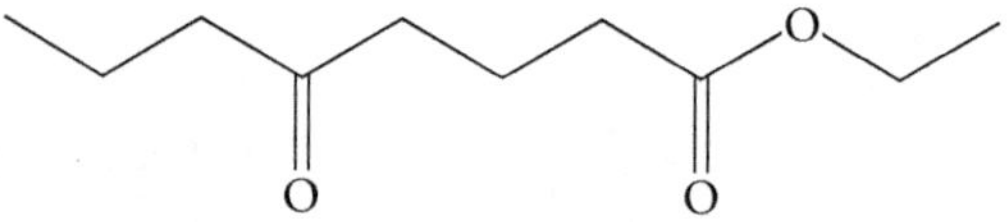 [structure]

(C) 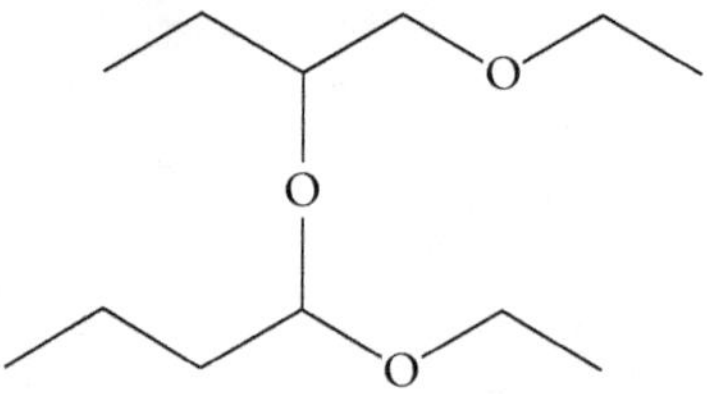 [structure]

(D) 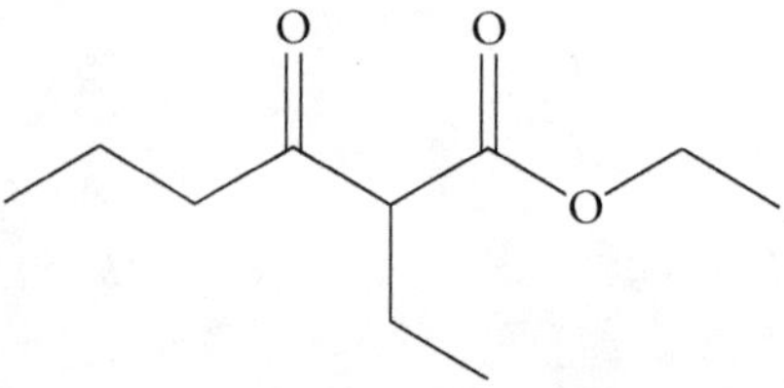 [structure]

301. An organic chemist was investigating the following reaction:

What was the structure of the major product?

(A)

(B)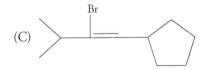

(C)

(D)

302. Which of the following is the most reactive?
(A) Acid chloride
(B) Acid anhydride
(C) Ether
(D) Ester

303. An organic chemist was investigating the following reaction of an alcohol with PCC (shown above the reaction arrow):

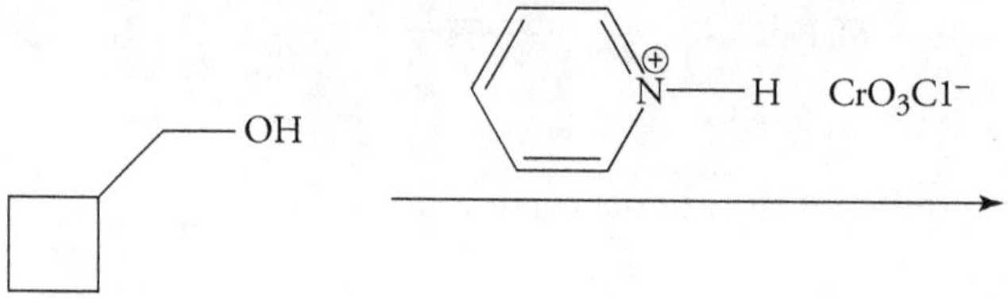

What was the structure of the major product?

(A)

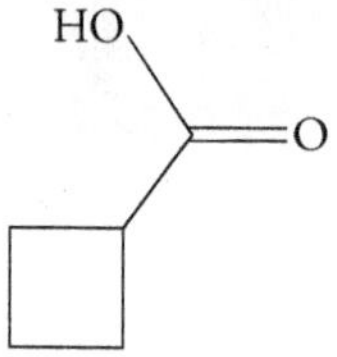

(B)

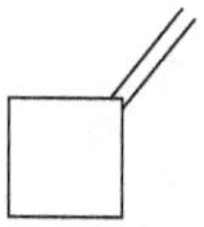

(C)

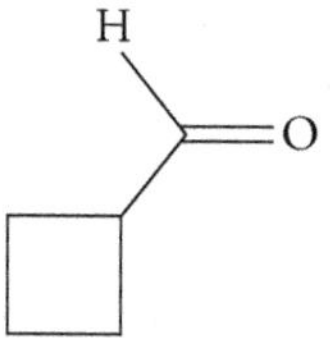

(D)

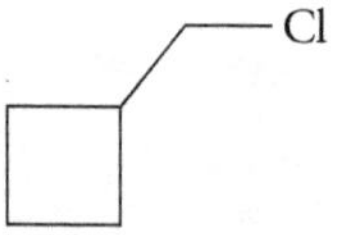

304. Which of the following compounds is the strongest acid?

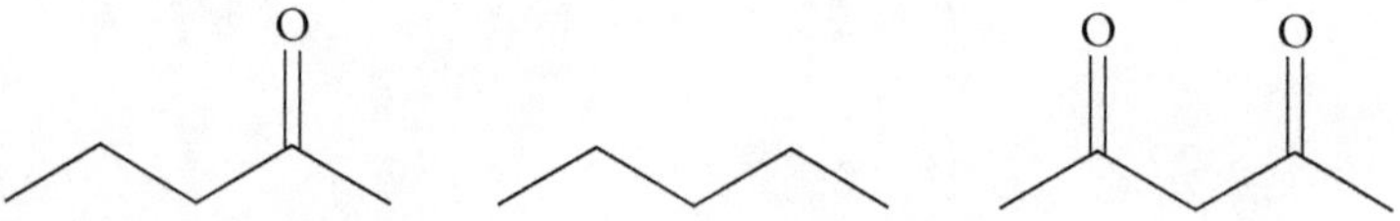

I II III

(A) III
(B) II
(C) I
(D) All are equally strong.

CHAPTER 6

Lab Technique and Spectroscopy

Questions 305–311 refer to the following passage.

There are a variety of instrumental methods that can give information about the structure of molecules. In general, the absorption of energy (and often the subsequent release of energy) gives information about the presence of some feature of the molecule. In infrared (IR) spectroscopy, the absorption of energy in the infrared region of the spectrum gives evidence concerning the types of bonds present. Ultraviolet-visible (UV-Vis) spectroscopy yields information about the molecular orbital arrangement in a molecule. Finally, nuclear magnetic resonance (NMR) spectroscopy utilizes the absorption of energy in the radio-wave portion of the spectrum to give information concerning the environments occupied by certain nuclei, especially ^{1}H and ^{13}C.

Light with energy in the infrared region of the electromagnetic spectrum has enough energy to cause a covalent bond to vibrate. If the vibration causes a change in the dipole moment of the molecule, there will be absorption of energy. (The vibrations are due to the stretching and bending of the bonds.) Most organic compounds have several absorptions in the 4000–600 cm^{-1} region of the spectrum. All of the absorptions give information about the structure of the molecule. The region below 1500 cm^{-1} is the fingerprint region of the IR spectrum.

Ultraviolet and visible spectroscopy (UV-Vis) is an analytical technique that is useful in the investigation of some organic molecules. Absorption of energy in this region of the electromagnetic spectrum can excite an electron from the ground state to an excited state. This is usually from the HOMO (highest occupied molecular orbital) to the LUMO (lowest unoccupied molecular orbital). This technique is particularly useful for compounds containing multiple bonds.

In an NMR spectrometer, the sample is placed in a large external magnetic field. This external field forces the nuclei to align themselves either with the field or against the field. Nuclei with one alignment can absorb energy and switch to the other alignment, and vice versa. This process is called spin flipping. When spin flipping occurs, the nucleus is in resonance. The energy required to induce this transition is in the radio-frequency region of the electromagnetic spectrum.

There are other magnetic fields influencing the nuclei besides the external magnetic field. The electrons in the molecule also have their own magnetic field. The field due to the electrons tends to oppose the external magnetic field, which results in electron shielding. The amount of shielding depends upon the number of electrons. The greater the electron shielding present, the lower the energy requirement for resonance is (resulting in a downfield shift). The position of the absorption is referred to as the chemical shift. For proton NMR, the chemical shift is normally in the 0–15 ppm region relative to the standard TMS (tetramethyl silane, $Si(CH_3)_4$). (The abbreviation ppm refers to parts per million.) For ^{13}C NMR, the chemical shift is normally in the 0–200 ppm region.

Chemically equivalent nuclei will absorb at the same energy. Consider, for example, the structure of ethanol. There are three distinct types of hydrogen atoms in this structure. In the proton NMR spectrum of ethanol, the three hydrogen atoms of the CH_3 group are chemically equivalent, as are the two hydrogen atoms of the CH_2 group, and both are different from the hydrogen atom attached to the oxygen.

305. The monobromination of methylpropane gives predominantly one product. How many ^1H NMR signals does the minor product of this reaction have?
- (A) 4
- (B) 3
- (C) 2
- (D) 1

306. Which of the following molecules contains only singlets in its proton NMR spectrum?

(A)

(B)

(C)

(D)

307. When isopropyl alcohol is heated with sulfuric acid, two products may result. Depending upon the conditions, diisopropyl ether or propene may form. Which of the following observations in the IR spectrum would indicate that diisopropyl ether was the product?

(A) The absence of an O–H stretch
(B) The absence of a C=C stretch
(C) An increase in the intensity of the C–H stretch
(D) An increase in the band near 1700 cm^{-1}

308. How many ^{1}H NMR signals would be present in each of the following compounds?

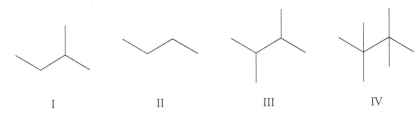

I   II   III   IV

(A) I has 4, II has 2, III has 2, and IV has 1.
(B) I has 5, II has 4, III has 1, and IV has 2.
(C) I has 5, II has 2, III has 3, and IV has 2.
(D) I has 3, II has 2, III has 1, and IV has 1.

309. Which of the following compounds will exhibit a band in the 1700 cm^{-1} region of its IR spectrum?

(A) $CH_3CH_2CH_2NH_2$
(B) $CH_3CH_2C{\equiv}CH$
(C) $CH_3CH_2CH_2OH$
(D) $CH_3CH_2CH{=}O$

310. An IR absorption near 3400 cm^{-1} might indicate the presence of:

(A) An aldehyde.
(B) An alkene.
(C) An alcohol.
(D) A ketone.

311. An unknown compound has a strong IR absorption near 1700 cm^{-1}. The proton NMR had no signal in the 9–10 ppm region and gave a negative Tollens' test. Iodoform forms when the compound reacts with a mixture of I_2/NaOH. Which of the following might the unknown compound be?

(A) 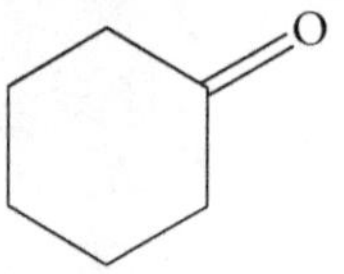 cyclohexanone

(B) 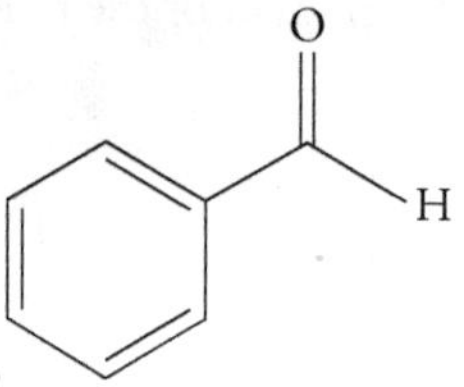 benzaldehyde

(C) 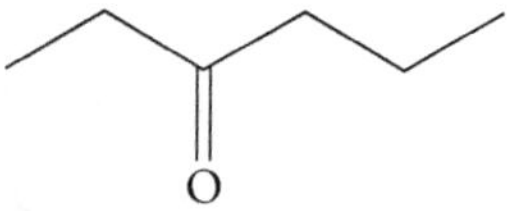 3-hexanone

(D) 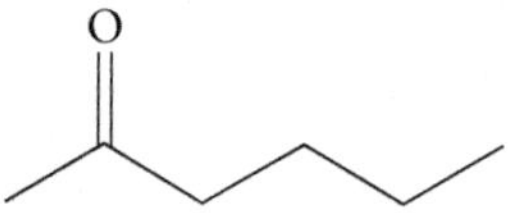 2-pentanone

312. The incomplete reduction of an ester with sodium borohydride may yield an alcohol and boric acid. A chemist attempts to separate the products by extraction. The first step in the extraction is to treat the mixture with aqueous sodium bicarbonate and extract with ether. The next step is to treat the ether layer with sodium hydroxide. What product(s) will be in the ether layer after treatment?

(A) Alcohol only
(B) Alcohol and ester
(C) Boric acid
(D) Boric acid and alcohol

313. How many signals would appear in the ^{13}C NMR spectrum of the following compound?

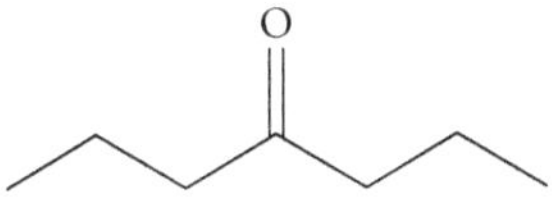

(A) 4
(B) 6
(C) 10
(D) 12

314. How many ^{13}C NMR resonances appear in the spectrum of the following compound?

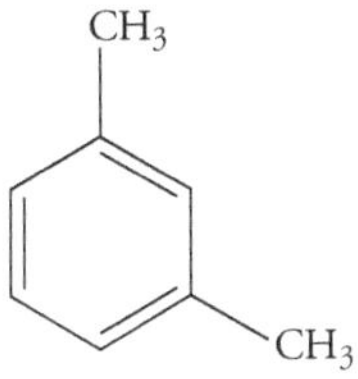

(A) 4
(B) 8
(C) 6
(D) 5

315. The following illustrates one method for synthesizing aspirin from salicylic acid and acetic acid:

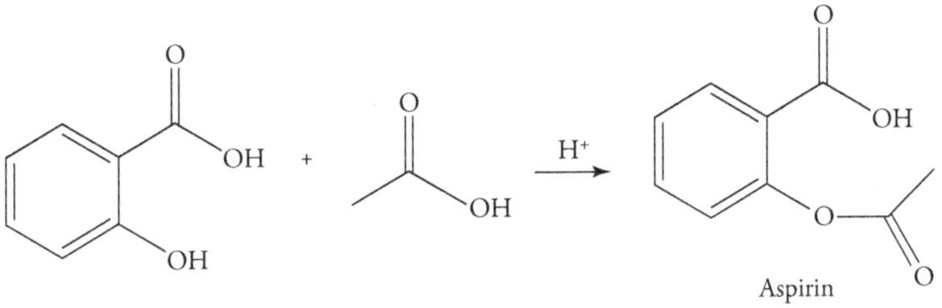

The reaction does not go to completion. A researcher attempts to separate the mixture with aqueous sodium bicarbonate. Which of the components of the mixture will be in the aqueous layer?

(A) Acetic acid and aspirin
(B) Salicylic acid and acetic acid
(C) Acetic acid only
(D) All three

316. An environmental chemist isolates a polluted sample that is believed to contain the following compounds:

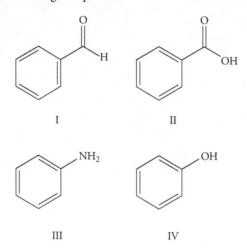

I II III IV

Extraction of the mixture with a sodium bicarbonate solution yielded an aqueous layer and an organic layer. Which of the compounds would be in the aqueous layer?

(A) II and IV
(B) I and III
(C) I and IV
(D) II and IV

317. One reaction of alkenes is ozonolysis. The reaction sequence begins with the reaction with ozone and ends with a treatment of the ozonide formed with zinc and water. What is the characteristic IR peak in every product?

(A) 1300 cm^{-1}
(B) 3300 cm^{-1}
(C) 1700 cm^{-1}
(D) 1250 cm^{-1}

318. Linolenic acid is an essential fatty acid that is required in the diet of humans. This acid is a polyunsaturated acid containing three double bonds and 18 carbon atoms. The systematic name is (9Z, 12Z, 15Z)-9,12,15-octadecatrienoic acid. Which of the isomers of linolenic acid will have the lowest melting point?

(A) (9Z, 12Z, 15Z)-9,12,15-octadecatrienoic acid
(B) (9Z, 12E, 15Z)-9,12,15-octadecatrienoic acid
(C) (9E, 12Z, 15E)-9,12,15-octadecatrienoic acid
(D) (9E, 12E, 15E)-9,12,15-octadecatrienoic acid

319. Which of the following laboratory techniques is NOT useful in the separation of a mixture of compounds?
    (A) Extraction
    (B) Chromatography
    (C) IR spectroscopy
    (D) Distillation

320. Two compounds with the formula C_3H_7ClO are 1-chloro-2-methoxyethane and 1-chloro-2-propanol. Which of the following laboratory techniques would be the simplest method for separating a mixture of these two compounds?
    (A) Extraction
    (B) Distillation
    (C) Thin-layer chromatography
    (D) NMR spectroscopy

321. A researcher needs to make a pH 9.2 buffer. The pK_a of the ammonium ion is 9.2. What is the mole ratio of ammonia to ammonium ion in this buffer?
    (A) 1.0
    (B) < 1.0
    (C) > 1.0
    (D) Impossible to determine without knowing the volume of the solution.

322. A chemist performs a reaction that yields a racemic mixture. What should she do next to separate the components of the mixture?
    (A) Crystallization
    (B) Fractionation
    (C) Resolution
    (D) Distillation

323. The monochlorination of an alkane gave three products. Which of the following would be the BEST method for determining the structures of the products?
    (A) Mass spectroscopy
    (B) NMR spectroscopy
    (C) IR spectroscopy
    (D) UV-Vis spectroscopy

324. Solvent extraction is a useful laboratory technique for separating mixtures. However, this method will not separate many combinations efficiently. What is a useful alternative to solvent extraction?

(A) Chromatography
(B) IR spectroscopy
(C) Ultracentrifugation
(D) Filtration

325. The mononitration of toluene yields a mixture of 2-nitrotoluene and 4-nitrotoluene. The formula of both compounds is $C_7H_7NO_2$. Which of the following would be the best way to distinguish these compounds?

(A) UV-Vis spectroscopy
(B) Mass spectroscopy
(C) IR spectroscopy
(D) NMR spectroscopy

326. The mononitration of toluene yields a mixture of 2-nitrotoluene and 4-nitrotoluene. The formula of both compounds is $C_7H_7NO_2$. Thin-layer chromatography (TLC) was used to monitor the progress of this reaction. On a normal silica gel TLC plate, the unreacted toluene traveled farther. Which of the following best explains why toluene traveled farther than either of the nitro products?

(A) Toluene has a lower molecular weight than nitrotoluene.
(B) Toluene is polar, and nitrotoluene is nonpolar.
(C) Toluene is nonpolar, and nitrotoluene is polar.
(D) Toluene chemically reacts with the silica, and nitrotoluene does not.

327. The exhaustive nitration of benzene yields a compound with the formula $C_6H_3N_3O_6$. How many signals appear in the ^{13}C NMR spectrum of 1,3,5-trinitrobenzene?

(A) 2
(B) 6
(C) 1
(D) 3

328. A chemist isolates the following three compounds from a reaction mixture:

Which of the following would be the best way to differentiate these compounds?
(A) UV-Vis spectroscopy
(B) Thin-layer chromatography
(C) Mass spectroscopy
(D) NMR spectroscopy

329. How many signals would appear in the ^{1}H NMR spectrum of the following compound?

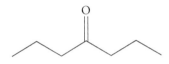

(A) 2
(B) 3
(C) 5
(D) 6

330. Compound Y has the formula $C_{10}H_{16}O_2$. Compound Y reacts with lithium aluminum hydride to form Compound Z, with the formula $C_{10}H_{20}O_2$. What might be the formula of the compound formed when Compound Z is heated with H_2SO_4?
(A) $C_{10}H_{16}$
(B) $C_{10}H_{16}O_3$
(C) $C_{10}H_{22}$
(D) $C_{10}H_8$

331. Which of the following compounds has three overlapping singlet peaks, each integrating for three hydrogen atoms? (This may appear as one singlet integrating for nine hydrogen atoms.)

(A) benzene with CH3 and CH3 (1,2-disubstituted)

(B) benzene with three CH3 groups

(C) benzene with CH3, OCH3, OCH3

(D) benzene with OCH3, OCH3, OCH3

332. The addition of sodium hydroxide to 1-bromopentane yields sodium bromide and 1-pentanol. Which of the two organic compounds (1-bromopentane or 1-pentanol) has the higher boiling point?
   (A) 1-pentanol because of hydrogen bonding
   (B) 1-bromopentane because of higher molecular weight
   (C) 1-pentanol because of polymerization
   (D) 1-bromopentane because it has a greater dipole moment

333. Which spectroscopic method affects the electronic structure of the molecule being studied?
   (A) UV-Vis spectroscopy
   (B) IR spectroscopy
   (C) Mass spectroscopy
   (D) NMR spectroscopy

334. Sulfonation of toluene will yield both ortho and para toluenesulfonic acid. In order to monitor the progress of the reaction, an organic chemist periodically removes a sample from the reaction mixture and places it on a normal silica gel TLC (thin-layer chromatography) plate. Which of the following is true?
    (A) Toluene travels farther because it is less polar then toluenesulfonic acid.
    (B) Toluene travels farther because it is more polar than toluenesulfonic acid.
    (C) Toluenesulfonic acid travels farther because it is less polar than toluene.
    (D) Toluenesulfonic acid travels farther because it is more polar than toluene.

335. The reduction of an ester with sodium borohydride can produce an alcohol and boric acid. A chemist attempts to separate the products by extraction. The first step in the extraction is to treat the mixture with aqueous sodium bicarbonate and extract with ether. What product(s) will be in the aqueous layer after treatment?
    (A) Boric acid only
    (B) Ester only
    (C) Boric acid and ester
    (D) All three products

336. How many ^{13}C resonances will appear in the NMR of the following compound?

    (A) 10
    (B) 8
    (C) 12
    (D) 14

337. The ^1H NMR spectrum of an organic compound showed, among other things, a triplet integrating for nine hydrogen atoms. How many hydrogen atoms are adjacent to the hydrogen atoms giving rise to this triplet?
    (A) 1
    (B) 2
    (C) 3
    (D) 9

338. The compound C_5H_8O has an IR band at 1715 cm^{-1}. What functional group is indicated?
   (A) A carboxylic acid
   (B) An ester
   (C) A ketone
   (D) An alcohol

339. Which of the following proton NMR signals represents the most deshielded proton?
   (A) 7.2 ppm
   (B) 9.8 ppm
   (C) 2.5 ppm
   (D) 1.5 ppm

340. The ^{1}H NMR of ketone shows five signals. Reduction of this compound with sodium borohydride and treating the product with acid followed by aqueous workup yields a new compound. How many new signals will be in the ^{1}H NMR of the new compound?
   (A) 0
   (B) 1
   (C) 2
   (D) 3

341. The major peaks in the IR spectrum of a compound are at 2990, 1715, 1640, and 915 cm^{-1}. Which functional group is NOT present?
   (A) Alcohol
   (B) Ester
   (C) Aldehyde
   (D) Ketone

342. The Kuhn-Winterstein reaction converts 1,2-glycols to trans alkenes using a variety of reagents, such as P_2I_4. The generic reaction is:

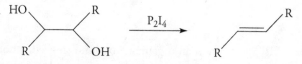

What is the easiest method of separating large amounts of the product from unreacted starting material?
   (A) Electrophoresis
   (B) Extraction into aqueous base
   (C) Thin-layer chromatography
   (D) Distillation

343. The Leuckart reaction involves the reaction of aldehydes or ketones with ammonium salts of formic acid. The general reaction is:

$$\underset{RR}{\overset{O}{\|}}{C} + NH_4^+CHO_2^- \xrightarrow{\Delta} \xrightarrow{H^+} \underset{RR}{\overset{NH_2}{|}}{C}$$

What is the easiest method of separating the product from unreacted starting material?

(A) First extract with water, then extract with acid.
(B) First extract with water, then extract with base.
(C) First extract with ether, then extract with alcohol.
(D) First extract with ether, then extract with base.

344. It is possible to separate amino acids by electrophoresis. During electrophoresis, an applied electric field separates the molecules. Every amino acid has at least two pK_a values, one for the carboxylic acid group and one for the amino group. Some amino acids have an additional pK_a value for the side chain. For alanine, the pK_a values are 2.3 and 9.6 (carboxylic acid and amino groups, respectively). Which of the following is true concerning an electrophoresis experiment involving alanine?

(A) In a 0.1 M NaOH solution, alanine will migrate toward the positive plate.
(B) In a 0.1 M NaOH solution, alanine will migrate toward the negative plate.
(C) In a 0.1 M HCl solution, alanine will migrate toward the positive plate.
(D) In a 0.1 M HCl solution, alanine will not migrate toward either plate.

345. A chemist extracts several compounds from a plant into ether. Through a series of extractions, she isolates a mixture of two compounds. Separation of the two compounds on a TLC plate showed Compound I to have an R_f value of 0.3 and Compound II to have an R_f value of 0.8. What do the R_f values indicate?

(A) Compound II is more polar than Compound I.
(B) Compound I is more polar than Compound II.
(C) Compound I has a higher molecular weight than Compound II.
(D) Compound II has a higher molecular weight than Compound I.

346. Compound I has the formula $C_{12}H_{18}O_2$ and shows an IR absorption at 1740 cm^{-1}. Treatment of the compound with aqueous HCl and heat resulted in another compound, II, with an IR absorption at 1710 cm^{-1}. What type of functional group was present in Compound I?
(A) Ketone
(B) Ether
(C) Carboxylic acid
(D) Ester

347. Compound X reacts with a mixture of I_2 in aqueous NaOH to form an odiferous yellow precipitate. What type of compound could Compound X be?
(A) A methyl ketone
(B) An alcohol
(C) An ether
(D) A carboxylic acid

348. Compound Y has the formula $C_{10}H_{16}O_2$. Compound Y reacts with lithium aluminum hydride to form Compound Z, with the formula $C_{10}H_{20}O_2$. How many rings could Compound Y have?
(A) 4
(B) 3
(C) 2
(D) 1

349. Benzoyl peroxide is useful in the formation of polymers such as polymethyl methacrylate (Lucite). This compound initiates polymerization following the homolytic cleavage of the O–O bond. What type of step is this?
(A) An initiation step
(B) A propagation step
(C) An instigation step
(D) A proliferation step

350. How many resonances would appear in the ^{13}C NMR of the following compound?

(A) 4
(B) 3
(C) 8
(D) 2

351. Which of the following compounds exhibits spin-spin coupling in its proton NMR?
(A) CH_3Cl
(B) $HOCH_2CH_2OH$
(C) $CH_3CH_2CH_2Cl$
(D) $CH_2=C=O$

352. Which of the following compounds will exhibit a band in the 2100 cm^{-1} region of its IR spectrum?
(A) $CH_3CH_2CH=O$
(B) $CH_3CH_2C\equiv CH$
(C) $CH_3CH_2CH_2OH$
(D) $CH_3CH_2CH_2NH_2$

353. What is the best treatment to extract an amine into an aqueous phase from a mixture?
(A) Use 0.1 M HCl.
(B) Use 0.1 M NaHCO3.
(C) Use 0.1 M NaOH.
(D) Use 0.1 M C_2H_5OH.

354. The monobromination of isobutane produces one major product. How many signals will there be in the proton NMR spectrum of the product?
(A) There will be a singlet and a multiplet.
(B) There will be only a singlet.
(C) There will be only a doublet.
(D) There will be a triplet and a quartet.

355. Which of the following would be the best method for separating 1-hexanol from dipropyl ether?
   (A) Filtration
   (B) Extraction with base
   (C) Extraction with acid
   (D) Fractional distillation

356. What splitting pattern is expected for the indicated methyl group in the following compound?

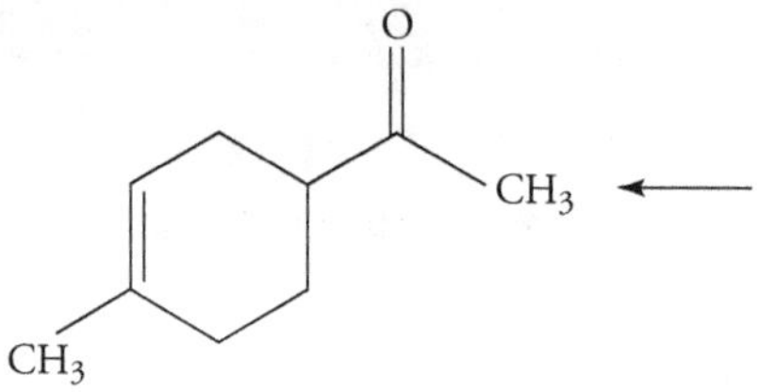

   (A) Singlet
   (B) Doublet
   (C) Triplet
   (D) Multiplet

357. A compound with the formula C_4H_8 shows only one peak in its NMR spectrum. What is the compound?
   (A) 2-Butene
   (B) Methylpropene
   (C) 1-Butene
   (D) Cyclobutane

358. How many ^{13}C NMR signals would be present in each of the following compounds?

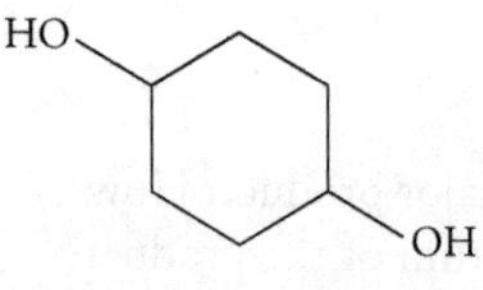

   I 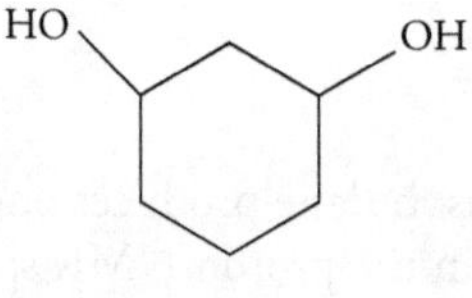   II 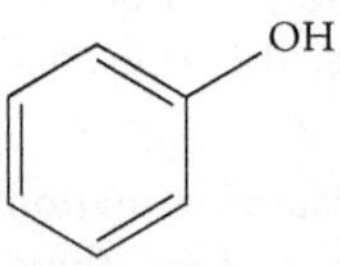   III    IV

   (A) I has 2, II has 4, III has 4, and IV has 1.
   (B) I has 6, II has 4, III has 2, and IV has 2.
   (C) I has 4, II has 3, III has 4, and IV has 2.
   (D) I has 2, II has 6, III has 3, and IV has 1.

359. An organic chemist investigates the following reaction:

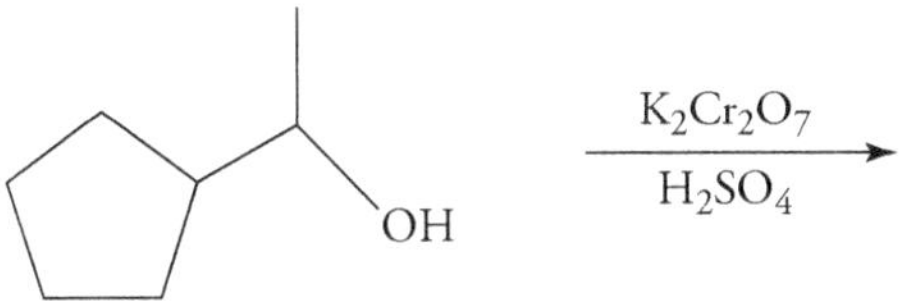

[cyclopentyl-CH(CH₃)-OH] → K₂Cr₂O₇ / H₂SO₄

Which of the following observations in the IR spectrum would indicate that the reaction went to completion?
(A) The disappearance of the C–C stretch
(B) The appearance of the C=O stretch
(C) The disappearance of the O–H stretch
(D) The appearance of the C=C stretch

360. An environmental chemist isolates a polluted sample believed to contain the following compounds:

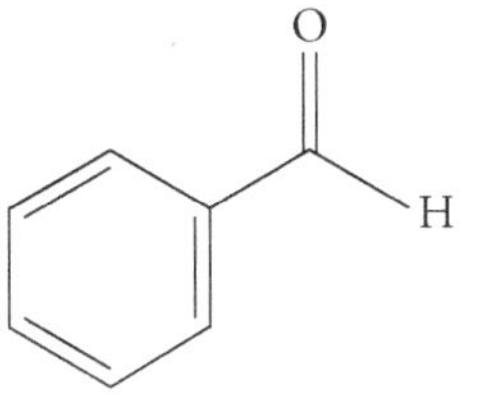

I: benzaldehyde
II: benzoic acid
III: aniline (PhNH₂)
IV: phenol (PhOH)

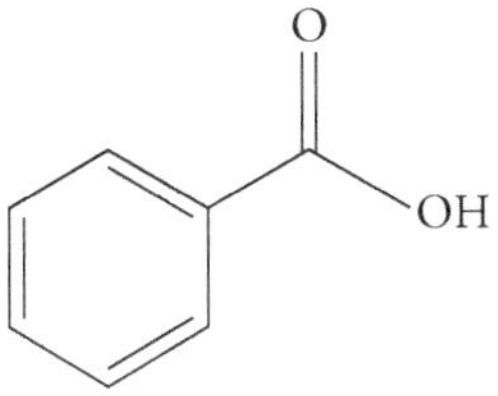

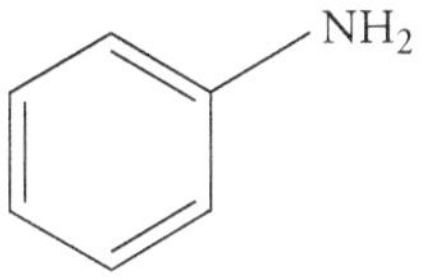

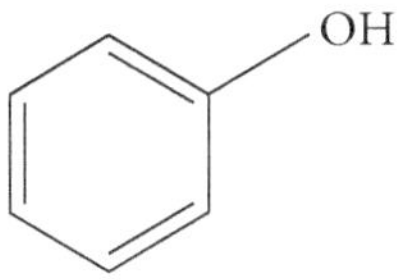

Extraction of the mixture with dilute HCl yielded an aqueous layer and an organic layer. Which of these compounds would be in the aqueous layer?
(A) III, I, and II
(B) II and IV
(C) IV
(D) III

361. The monobromination of 2-methylbutane can yield a number of products. What would be the most efficient method of separating a 100-mL mixture of the isomers?

(A) Fractional distillation
(B) Solvent extraction
(C) Paper chromatography
(D) Titration

362. The following illustrates one method for synthesizing aspirin from salicylic acid and acetic acid.

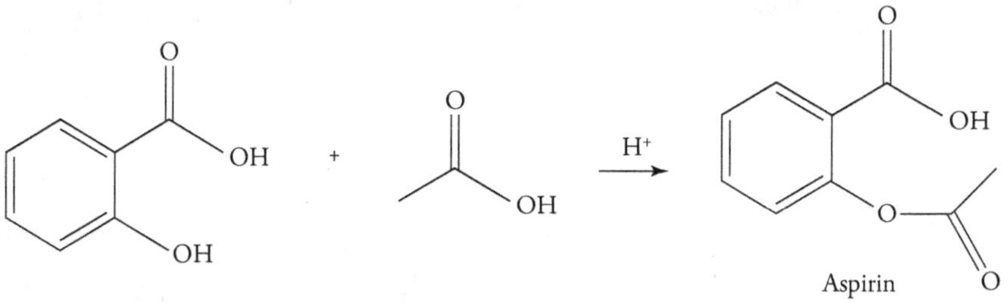

Which of the compounds will show a strong absorption near 1700 cm^{-1} in the IR spectrum?

(A) Aspirin only
(B) Salicylic acid and acetic acid
(C) Acetic acid only
(D) All three

363. Epoxides react with methyl sulfide ions via an S$_N$2 mechanism. The following is a typical example:

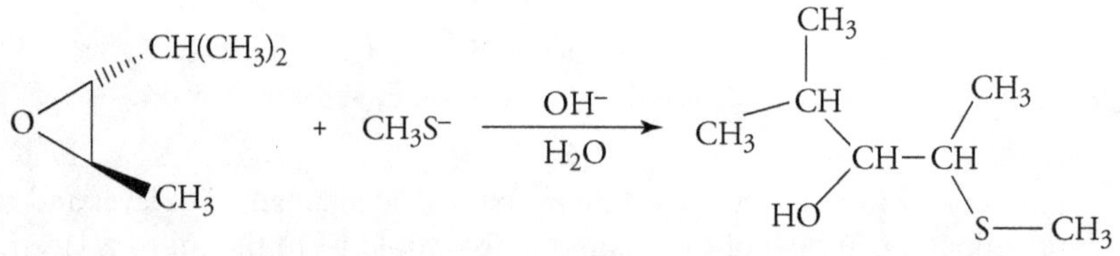

It would be possible to monitor the progress of this reaction by removing samples, neutralizing, extracting to separate the material from water, and then examining changes in which region of the IR spectrum?

(A) 1700 cm^{-1}
(B) 3000 cm^{-1}
(C) 3600 cm^{-1}
(D) 1000 cm^{-1}

364. The free radical monochlorination of 2-methylbutane will give the following products:

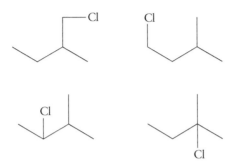

Which of the following methods would be the best at identifying the different products?
(A) ^{1}H NMR spectroscopy
(B) IR spectroscopy
(C) Mass spectroscopy
(D) UV-Vis spectroscopy

365. How many ^{1}H NMR resonances appear in the aromatic region of the following ion?

(A) 0
(B) 3
(C) 5
(D) 4

366. It is possible to separate amino acids by electrophoresis. During electrophoresis, an applied electric field separates the molecules. Every amino acid has at least two pK_a values, one for the carboxylic acid group and one for the amino group. Some amino acids have an additional pK_a value for the side chain. For asparagine, the pK_a values are 2.0 and 8.8 (carboxylic acid and amino groups, respectively). The pI value for an amino acid is the average of the pK_a values. At a pH = pI, the carboxylic acid group is deprotonated and the amino group is protonated. Which of the following is true concerning an electrophoresis experiment?

(A) At pH = 5.4, asparagine will not migrate toward either plate.
(B) At pH = 5.4, asparagine will migrate toward the positive plate.
(C) At pH = 1.0, asparagine will migrate toward the positive plate.
(D) At pH = 11.0, asparagine will migrate toward the negative plate.

CHAPTER 7

Bioorganic Chemistry

Questions 367–371 refer to the following figure.

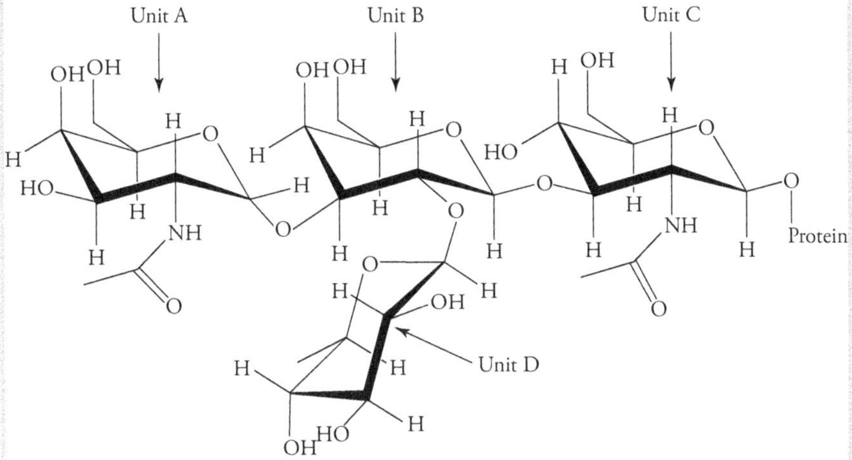

This carbohydrate sequence represents the terminal sequence of the Type A determinant protein in blood.

367. Which unit will give a positive Benedict's test as a reducing sugar?
- (A) None
- (B) Unit A
- (C) Unit C
- (D) Unit D

368. What are the forms of the anomeric carbon atoms in units A and D?
- (A) A is an acetal, and D is a hemiacetal.
- (B) A is a hemiacetal, and D is an acetal.
- (C) A is a hemiacetal, and D is a hemiacetal.
- (D) A is an acetal, and D is an acetal.

369. How many degrees of unsaturation are present in this carbohydrate sequence?
   (A) 8
   (B) 4
   (C) 9
   (D) 6

370. Which unit is a maltose unit?
   (A) A
   (B) B
   (C) D
   (D) None

371. What is the best description of the linkage between unit D and unit B?
   (A) β-1,3
   (B) β-1,2
   (C) α-1,2
   (D) α-1,3

Questions 372–374 refer to the following passage.

The partial hydrolysis of a hexapeptide yielded the following dipeptide fragments:

gly-ala   ser-gly   ser-pro   ala-ser   gly-ser

The following tripeptides were also found:

   gly-ala-ser   gly-ser-pro   ala-ser-gly

Analysis of the hexapeptide found that it contained 1 ala residue, 2 gly residues, 1 pro residue, and 2 ser residues. Treatment of the hexapeptide with 2,4-dinitrofluorobenzene followed by hydrolysis gave only residues with the dinitrophenyl group attached to glycine.

372. This information indicates that 2,4-dinitrofluorobenzene is useful in protein analysis because:
   (A) It is the most abundant amino acid in the protein.
   (B) It attaches to the C-terminal amino acid.
   (C) It attaches to the N-terminal amino acid.
   (D) It is the most reactive amino acid in the protein.

373. What is the amino acid sequence?
   (A) gly-ala-ser-gly-ser-pro
   (B) gly-ser-pro-gly-ala-ser
   (C) ala-ser-gly-gly-ser-pro
   (D) pro-ser-gly-ser-ala-gly

374. Treatment of an octapeptide with 2,4-dinitrofluorobenzene followed by hydrolysis gave only residues with the dinitrophenyl group attached to glycine. What was the reaction mechanism leading to the attachment of the 2,4-dinitrofluorobenzene?
   (A) Nucleophilic aromatic substitution
   (B) Electrophilic aromatic substitution
   (C) S_N1
   (D) E2

375. This is the structure of nicotine:

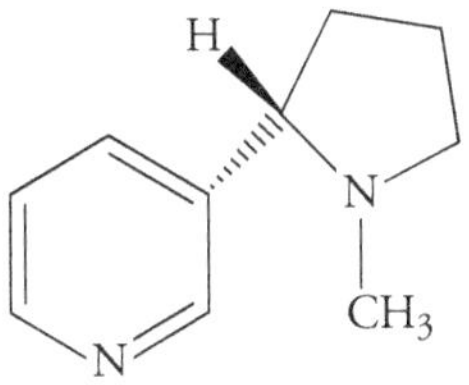

   What is the absolute configuration of nicotine?
   (A) R
   (B) S
   (C) Achiral
   (D) Impossible to determine

376. The generic formula for an amino acid is:

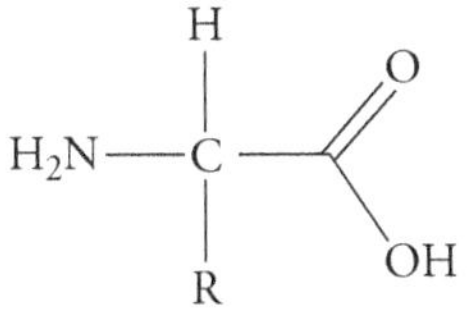

   Why are amino acids considered to be amphoteric?
   (A) They have both D and L forms.
   (B) They may undergo oxidation or reduction.
   (C) They may behave as an acid or as a base.
   (D) They can form enantiomers.

377. Maltose is a disaccharide formed by joining two glucose molecules by a glycoside linkage. The structure of maltose is:

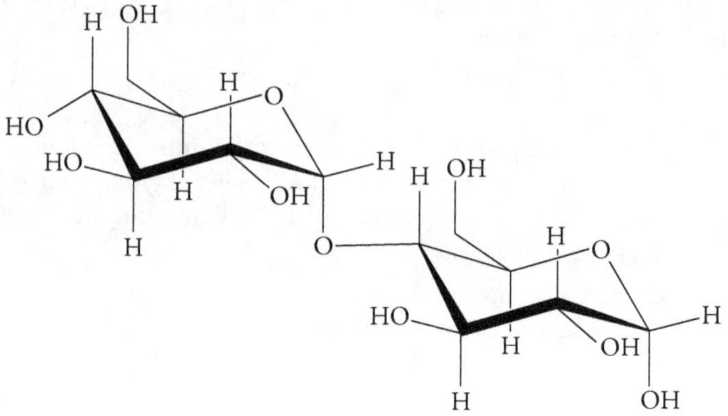

Which of the following best describes the glycoside linkage?
(A) α-1,4
(B) β-1,4
(C) α-1,2
(D) β-1,2

378. The disaccharide lactose has the following structure:

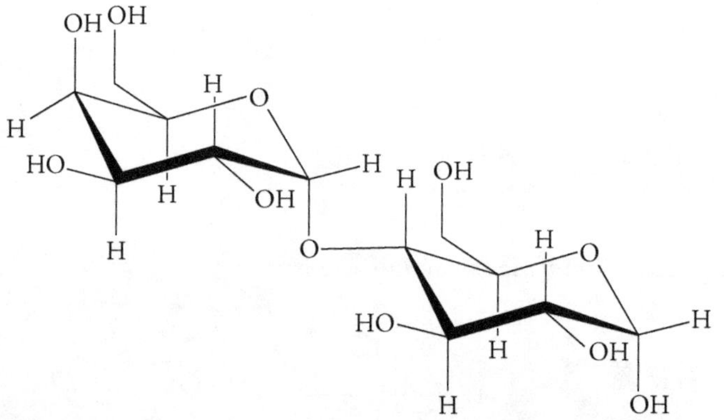

Which of the following is NOT true about lactose?
(A) It has a hemiacetal.
(B) It has more than 100 stereoisomers.
(C) It is a dipeptide.
(D) It has a glycoside linkage.

379. The structure of testosterone is:

How many stereoisomers of testosterone are possible?
(A) 6
(B) 24
(C) 32
(D) 64

380. The anomers α-D-ribose and β-D-ribose differ only in the stereochemistry about the anomeric carbon atom. What is the term referring to the conversion of the α-anomer to the β-anomer?
(A) Isomerization
(B) Mutarotation
(C) Anomerization
(D) Tautomerism

381. Physiological pH is 7.4. What does this mean?
(A) It is slightly more basic than pure water.
(B) It is slightly more acidic than pure water.
(C) The concentration of H^+ is less than the concentration of OH^-.
(D) Both A and C.

382. The pI of the amino acid leucine is 5.98. A sample of leucine is placed in a solution buffered at a pH of 5.98. Then two electrodes are placed into the solution and connected to a battery. Toward which electrode will the majority of the leucine migrate?
(A) Toward the negative electrode
(B) Toward the positive electrode
(C) Toward neither electrode
(D) Half toward each electrode

383. Which of the following sugars are aldopentoses?

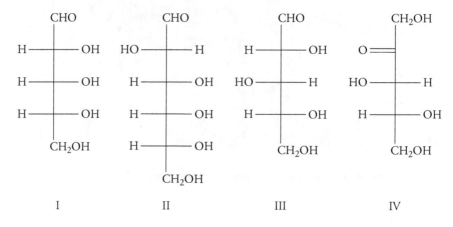

(A) I and III
(B) I, III, and IV
(C) I, II, and III
(D) II and IV

384. Pentoses and hexoses may form cyclic structures with five or six atoms in the ring. Why do they not form cyclic structures with smaller ring sizes?
(A) Smaller rings rapidly polymerize to form polysaccharides.
(B) Smaller rings require unstable angles that are greater than tetrahedral.
(C) Smaller rings rapidly react to form disaccharides.
(D) Smaller rings require unstable angles that are less than tetrahedral.

385. Which of the following is normally characteristic of the positioning of the polar side chains in the tertiary structure of proteins?
(A) They are in the active site.
(B) They are in position to react with each other.
(C) They are on the exterior of the protein.
(D) They are in the interior of the protein.

386. What type of compound is the phosphate ester of a nucleoside?
(A) DNA
(B) Nucleotide
(C) Pyramidine
(D) RNA

387. Many organic functional groups are important in biological systems. For example, fats contain ester groups, and proteins contain amide groups. In the laboratory, both acid chlorides and acid anhydrides are useful in the synthesis of these functional groups. However, neither acid chlorides nor acid anhydrides are important in the biosynthesis of esters or amides. Why are acid chlorides and acid anhydrides NOT important in biological systems?
    (A) They are too reactive to exist in biological systems.
    (B) They are nonpolar and are not soluble in biological (aqueous) systems.
    (C) They react only at higher temperatures than are present in biological systems.
    (D) They are less efficient than the enzymes that are present in biological systems.

388. Saponification is the base-catalyzed hydrolysis of an ester. The saponification of a fat produces glycerol and the carboxylate salts of the base. The carboxylate portion typically has 10 to 20 carbon atoms. In aqueous solution, the carboxylate salt exists in a structure known as a micelle. Which of the following best describes the structure of a micelle?
    (A) The conglomerate has hydrophilic carboxylate groups on the exterior and the hydrophobic hydrocarbon portions on the interior.
    (B) The conglomerate has hydrophobic carboxylate groups on the exterior and the hydrophilic hydrocarbon portions on the interior.
    (C) The conglomerate has hydrophilic carboxylate groups on the interior and the hydrophobic hydrocarbon portions on the exterior.
    (D) The conglomerate has hydrophobic carboxylate groups on the interior and the hydrophilic hydrocarbon portions on the exterior.

389. One of the many methods for synthesizing amino acids in the laboratory is the Strecker synthesis. In one experiment, an organic chemist utilized the method to prepare an amino acid. The compound she isolated was not optically active. Which of the following is the isolated amino acid?
    (A) Alanine
    (B) Proline
    (C) Glycine
    (D) Lysine

390. The structure of the amino acid ornithine is:

$$H_2N-CH(CH_2CH_2CH_2NH_2)-C(=O)-OH$$

How does the isoelectric point of alanine compare to that of ornithine?
(A) Alanine has a higher value.
(B) Alanine has a lower value.
(C) Alanine has about the same value.
(D) It is not possible to determine from the information given.

391. It is possible to synthesize an amino acid from an aldehyde through the Strecker synthesis. The general method for synthesizing an amino acid is:

$$R-CHO \xrightarrow[HCN]{NH_3} R-CH(NH_3^+)-CN \xrightarrow[Heat]{H^+/H_2O} R-CH(NH_3^+)-COO^-$$

What is the role of the ammonia?
(A) It is a nucleophile.
(B) It is an electrophile.
(C) It is an acid.
(D) It is a catalyst.

392. The structures of the amino acids serine and lysine are:

Serine: $H_2N-CH(CH_2OH)-C(=O)-OH$

Lysine: $H_2N-CH(CH_2CH_2CH_2CH_2NH_2)-C(=O)-OH$

Which of the following statements is true?
(A) Serine has two pK_a values, and lysine has two pK_a values.
(B) Serine has three pK_a values, and lysine has three pK_a values.
(C) Serine has one pK_a value, and lysine has two pK_a values.
(D) Serine has two pK_a values, and lysine has three pK_a values.

393. What is the most common laboratory method for the complete hydrolysis of a protein?
(A) Add hot concentrated base.
(B) Add hot concentrated acid.
(C) Add dilute acid at room temperature.
(D) Add dilute base at room temperature.

394. Alitame is an artificial sweetener with the following structure:

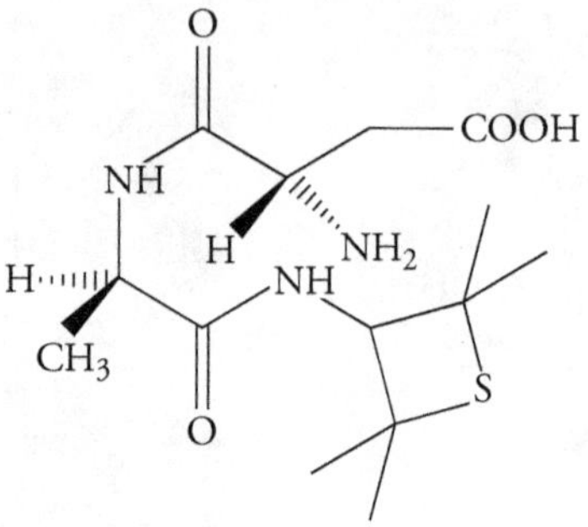

Alitame is more stable than aspartame and about 2,000 times as sweet as sucrose. What is the most basic group in alitame?

(A) The amino group
(B) Both the amide nitrogen atoms
(C) The sulfur
(D) The carbonyl oxygen atoms

395. Para-aminosalicyclic acid is a secondary line drug for the treatment of tuberculosis. The structure of para-aminosalicyclic acid is:

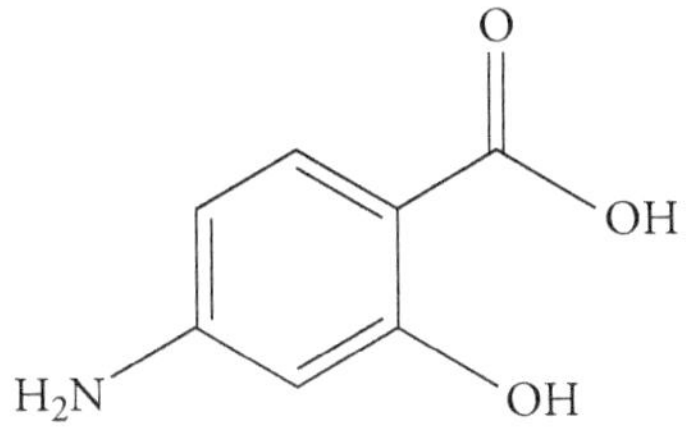

What is the form of this drug at pH = 14?

(A)

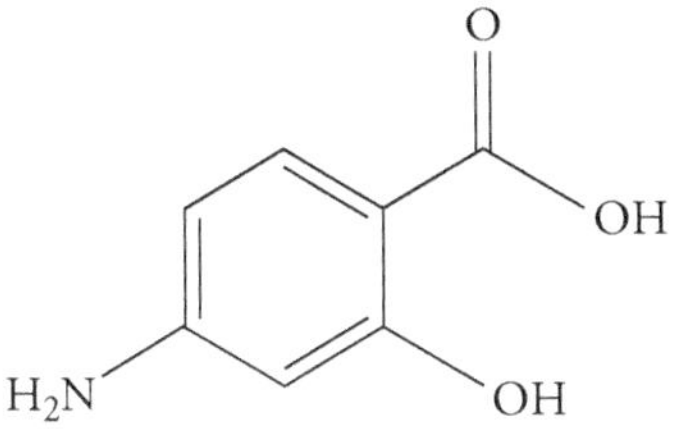

(B)

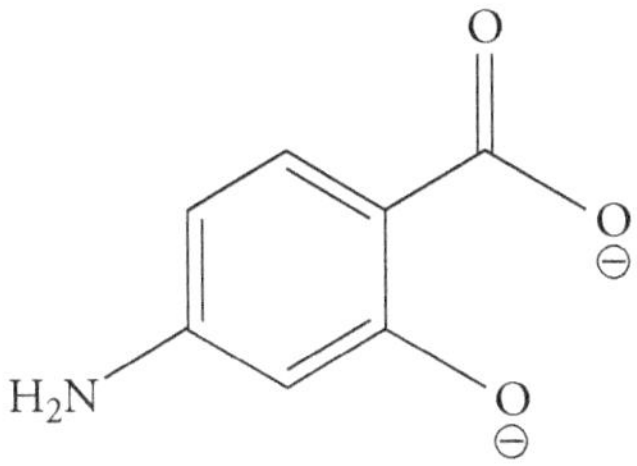

(C)

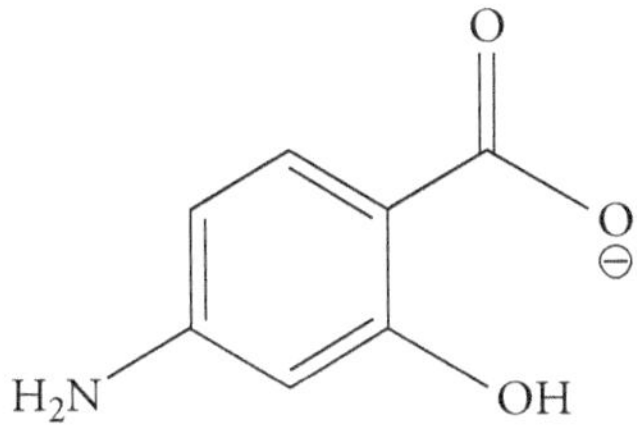

(D)

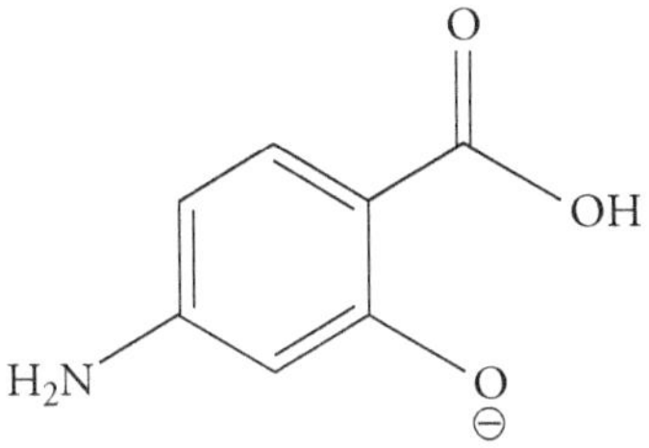

396. Reduction of the aldehyde group of the carbohydrate D-allose yields allitol. The structure of allitol is:

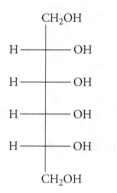

Which of the following statements best describes allitol?
(A) It is a chiral compound.
(B) It is a carbohydrate.
(C) It is a meso compound.
(D) It is a polymeric compound.

397. Referring to the figure in question 396, what type of compound would form if the alcohol group on the second carbon atom underwent oxidation?
(A) A ketohexose
(B) An aldohexose
(C) An acidic alcohol
(D) A carboxylic acid

398. Which of the following configurations may be present in naturally occurring carbohydrates?
I α   II β   III D   IV L
(A) I and III
(B) I, II, and III
(C) I, II, and IV
(D) I, II, III, and IV

399. This is the structure of the amino acid isoleucine:

$$H_2N-CH(CH(CH_3)CH_2CH_3)-C(=O)-OH$$

Which of the following is the zwitterion form of isoleucine?

(A) $H_2N-CH(CH(CH_3)CH_2CH_3)-C(=O)-OH$

(B) $H_2N-CH(CH(CH_3)CH_2CH_3)-C(=O)-O^{\ominus}$

(C) $H_3\overset{\oplus}{N}-CH(CH(CH_3)CH_2CH_3)-C(=O)-OH$

(D) $H_3\overset{\oplus}{N}-CH(CH(CH_3)CH_2CH_3)-C(=O)-O^{\ominus}$

400. Which of the following Fischer projections represents an L-monosaccharide?

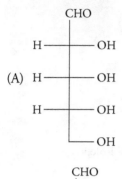

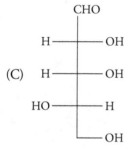

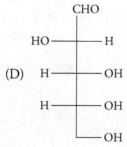

401. Alitame is an artificial sweetener with the following structure:

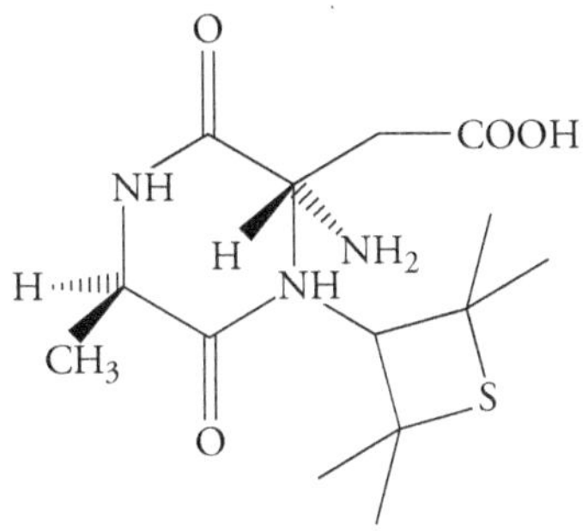

Alitame is more stable than aspartame and about 2,000 times as sweet as sucrose. Which of the following is true about alitame in the stomach (pH = 1)?

(A) The amino group would be protonated.
(B) The amide nitrogen atoms would be protonated.
(C) All the nitrogen atoms would be protonated.
(D) The overall charge would be 0.

402. Most amino acids are chiral compounds. Which of the following amino acids is achiral?

(A) $H_2N-CH(CH_2SH)-C(=O)-OH$

(B) $H_2N-CH(H)-C(=O)-OH$

(C) $H_2N-CH(CH(OH)CH_3)-C(=O)-OH$

(D) proline (pyrrolidine-2-carboxylic acid)

403. The following presents two representations of D-gulose. Which position in the right-hand structure is the same as the circled alcohol group in the left-hand structure?

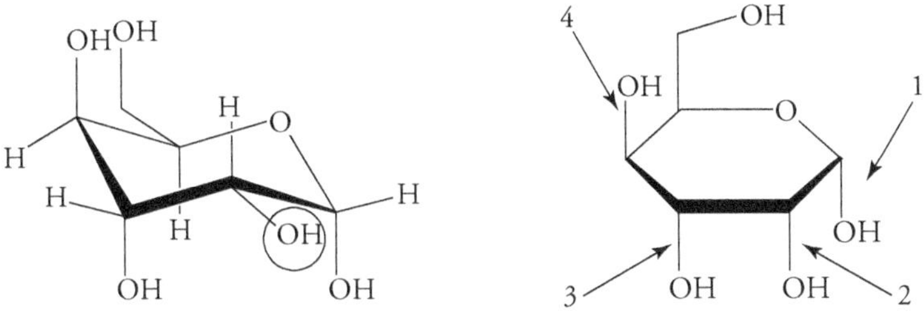

(A) 6
(B) 4
(C) 3
(D) 2

404. An amino acid is at its isoelectric point. Decreasing the hydrogen ion concentration will cause what change in the overall charge?

(A) Neutral to positive
(B) Neutral to negative
(C) Positive to negative
(D) Negative to positive

405. Which of the following amino acids would most likely be in a nonpolar region of a protein?

(A)
$$H_2N-CH(CH_2C_6H_5)-C(=O)-OH$$

(B)
$$H_2N-CH(CH_2SH)-C(=O)-OH$$

(C)
$$H_2N-CH(CH(OH)CH_3)-C(=O)-OH$$

(D)
$$H_2N-CH(CH_2OH)-C(=O)-OH$$

406. The drug thalidomide was introduced in the late 1950s to alleviate the morning sickness that often accompanies pregnancy. Synthesis of the drug resulted in a racemic mixture. One of the enantiomers was the active ingredient for treating morning sickness, while the other enantiomer led to birth defects. The following is one of the enantiomers of thalidomide. What is the most likely explanation of why the two forms of thalidomide have different activities?

(A) The interaction with a receptor or enzyme
(B) The ability to transport into the cells
(C) The greater solubility of one form
(D) The greater polarity of one form

407. Which of the following would be an example of an α-amino acid?

(A) [proline structure: pyrrolidine ring with HN and C(=O)—OH substituent]

(B) H$_2$N—CH$_2$—CH$_2$—C(=O)—OH

(C) H$_2$N—CH(CH$_3$)—C(=O)—NH$_2$

(D) H$_2$N—CH(CH$_2$SH)—C(=O)—OMe

408. Alitame is an artificial sweetener with the following structure:

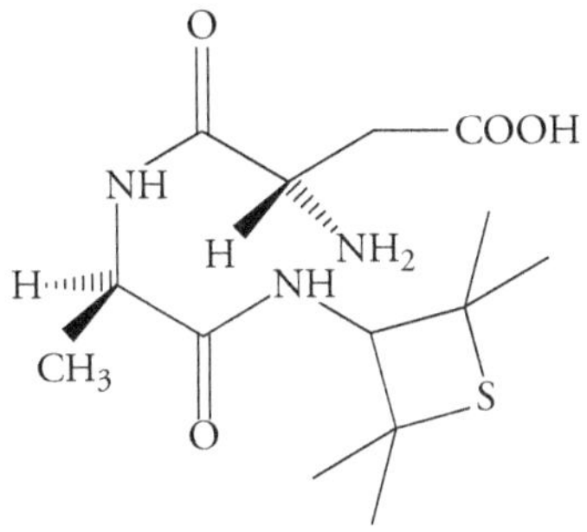

Alitame is more stable than aspartame and about 2,000 times as sweet as sucrose. What would the ^{1}H NMR of the methyl group attached to the chiral carbon appear as?

(A) It would appear as a singlet.
(B) It would appear as a triplet.
(C) It would appear as a doublet.
(D) It would appear as a quartet.

409. The structure of the amino acid ornithine is:

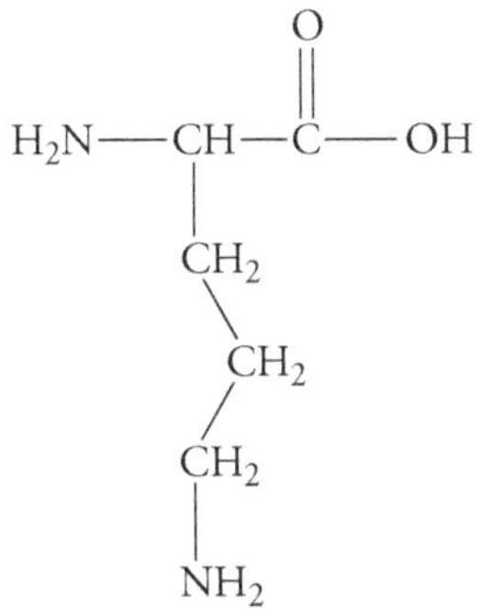

At what pH is this the predominant form of ornithine?

(A) It is stable near a neutral pH.
(B) It is stable at low pHs.
(C) It is stable at high pHs.
(D) It is never the predominant form.

410. Homoserine is not a common amino acid, but it is important in the biosynthesis of methionine, threonine, and isoleucine. Which of the following is the best representation of the zwitterionic form of homoserine?

(A)
$$H_2N-CH(CH_2CH_2OH)-C(=O)-OH$$

(B)
$$H_3\overset{\oplus}{N}-CH(CH_2CH_2OH)-C(=O)-OH$$

(C)
$$H_2N-CH(CH_2CH_2OH)-C(=O)-O^{\ominus}$$

(D)
$$H_3\overset{\oplus}{N}-CH(CH_2CH_2OH)-C(=O)-O^{\ominus}$$

411. The amino acid valine has a pK_a = 2.3 for the carboxyl end and a pK_a = 9.6 for the amino end. This means that the pI for valine is 5.95 (the average of the two pK_a values). Which of the following is true concerning a 1 mM valine solution buffered at pH 2.3?

(A) There are equal concentrations of the zwitterion form and the form with a +1 net charge.
(B) There are equal concentrations of the zwitterion form and the form with a −1 net charge.
(C) Only the fully protonated form is present.
(D) Only the fully deprotonated form is present.

412. The amino acid valine has a pK_a = 2.3 for the carboxyl end and a pK_a = 9.6 for the amino end. This means that the pI for valine is 5.95 (the average of the two pK_a values). Which of the following is true concerning a 1 mM valine solution buffered at pH 12?

(A)
$$H_3\overset{+}{N}-CH(CH(CH_3)CH_3)-C(=O)-O^-$$

(B)
$$H_2N-CH(CH(CH_3)CH_3)-C(=O)-OH$$

(C)
$$H_2N-CH(CH(CH_3)CH_3)-C(=O)-O^-$$

(D)
$$H_3\overset{+}{N}-CH(CH(CH_3)CH_3)-C(=O)-OH$$

413. Enzymes known as α-glucosidases catalyze the hydrolysis of carbohydrates to monosaccharides. Two of these enzymes are sucrase and maltase. These enzymes hydrolyze sucrose and maltose, respectively. The action of these enzymes produces which of the following monosaccharides?
    (A) Glucose and fructose
    (B) Glucose and sorbose
    (C) Sorbose and fructose
    (D) Glucose only

414. Examine the following two carbohydrates:

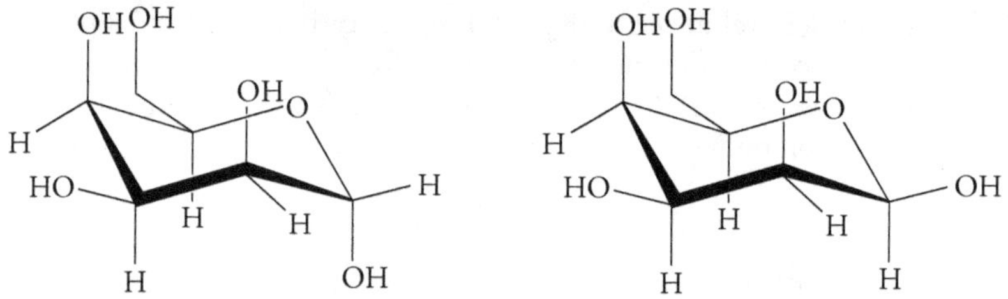

    Which of the following terms would be useful in describing the relationship between these molecules?
    I. Epimers
    II. Anomers
    III. Diastereomers
    (A) I, II, and III
    (B) I and II
    (C) II only
    (D) I only

415. The compound miglitol is an inhibitor of α-glucosidases. The structure of miglitol is:

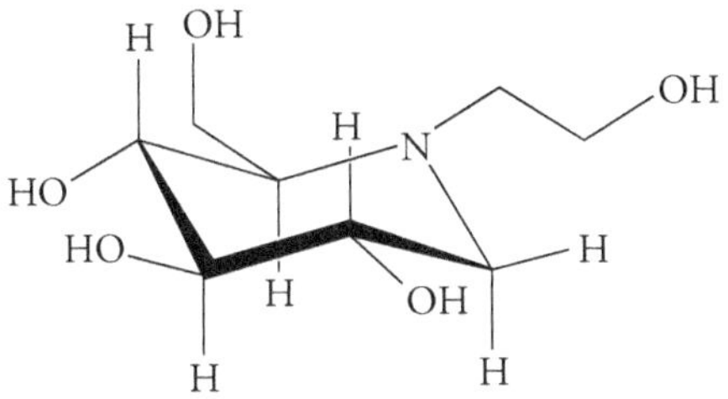

What are the absolute configurations about carbon atoms 2 and 5?
(A) 2S and 5R
(B) 2R and 5R
(C) 2S and 5S
(D) 2R and 5S

416. The following reaction, accompanied by ATP and appropriate enzymes, occurs in biological systems:

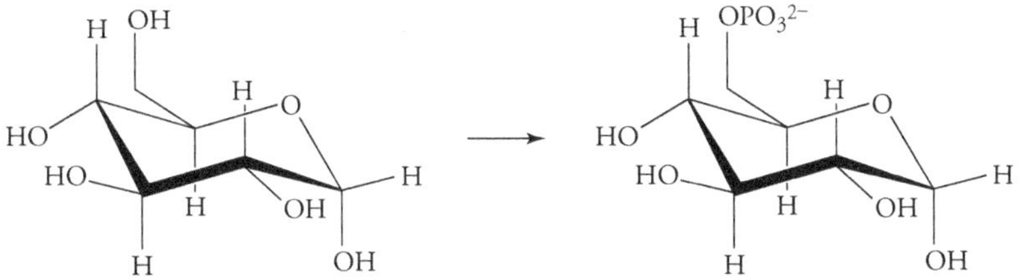

What functional group forms during this reaction?
(A) An ether
(B) An acid anhydride
(C) An ester
(D) A phosphine

417. Epimerase is an enzyme that can catalyze the isomerization of D-ribulose 5-phosphate to D-xylulose 5-phosphate. Which of the following is NOT an isomerization reaction?

(A) $CH_3CH_2COCl + H_2O \rightarrow HCl + CH_3CH_2COOH$
(B) trans-$CH_3CH=CHCH_3 \rightarrow$ cis-$CH_3CH=CHCH_3$
(C) $CH_3CH_2C(CH_3)_3 \rightarrow (CH_3)_2CHCH(CH_3)_2$
(D) $CH_3CH_2CHOHCHOHCH_3 \rightarrow CH_3CH_2C(OH)_2CH_2CH_3$

418. Which of the following is true concerning the classification of monosaccharides with six carbon atoms?
    (A) D-glucose and D-fructose have the same number of stereoisomers.
    (B) Each D-aldose has two anomers.
    (C) D-glucose has 32 stereoisomers.
    (D) Aldoses and hexoses have the same number of stereoisomers.

419. Polyunsaturated fats have lower melting points than saturated fats of comparable size. The unsaturation in most natural polyunsaturated fats is due to carbon-carbon double bonds, which are present exclusively as the cis isomer. Industrially, the partial hydrogenation of a polyunsaturated fat is a useful technique to increase the melting point and yield a commercial product. What conditions are necessary for partial hydrogenation of a fat?
    (A) $LiAlH_4/Et_2O$
    (B) H^+/H_2O
    (C) H_2/Ni
    (D) H_2/Δ

420. Which part of the structure of a protein is the result of the formation of hydrogen bonds between the peptide bonds of a protein?
    (A) Primary structure
    (B) Secondary structure
    (C) Tertiary structure
    (D) Quaternary structure

421. Peptide bonds hold the amino acids in a protein together. What is another name for a peptide bond?
    (A) Hydrogen bond
    (B) Glycoside linkage
    (C) Ester bond
    (D) Amide bond

422. Polyunsaturated fats have lower melting points than saturated fats of comparable size. The unsaturation in most natural polyunsaturated fats is due to carbon-carbon double bonds, which are present exclusively as the cis isomer. Industrially, the partial hydrogenation of a polyunsaturated fat is a useful technique to increase the melting point and yield a commercial product. The increase in the melting point is a result of increased:
    (A) Hydrogen bonding between the straighter chains.
    (B) London dispersion forces between the straighter chains.
    (C) London dispersion forces due to isomerization of the double bonds.
    (D) Hydrogen bonding due to isomerization of the double bonds.

423. Lactic acid has a pK_a of 3.85. What is the ratio of lactate ion to lactic acid in a solution at pH = 3.85?
(A) 1.00
(B) 3.85
(C) > 3.85
(D) 7.00

424. The zwitterion form of an amino acid is the predominant species in solution when the pH of the solution equals the pI of the amino acid. The pK_a of the carboxylic acid of alanine is 2.35, and the pK_a of the ammonium form of the amino group is 9.69. What is the pI value of alanine?
(A) 7.00
(B) 6.02
(C) 2.35
(D) 9.69

425. Acid hydrolysis of a tripeptide yielded equal amounts of each of the following three amino acids: leu, cys, and ala. What was the probable order of the amino acids in the tripeptide?
(A) ala-leu-cys
(B) leu-cys-ala
(C) leu-ala-cys
(D) It is impossible to determine.

426. Two molecules of the amino acid cysteine can join to form cystine. The reaction is:

$$H_2N-CH(CH_2SH)-COOH + HOOC-CH(CH_2SH)-NH_2 \rightleftharpoons H_2N-CH(CH_2-S-S-CH_2)-COOH \cdots$$

What type of reaction is this?
(A) Oxidation-reduction
(B) Neutralization
(C) Decomposition
(D) Polymerization

427. Glucose is a reducing sugar, which means that it will give a positive Benedict's test. A positive Benedict's test is the formation of a red precipitate of Cu_2O, which forms as the sugar undergoes oxidation by Cu^{2+}. The following equilibrium is present in all glucose solutions:

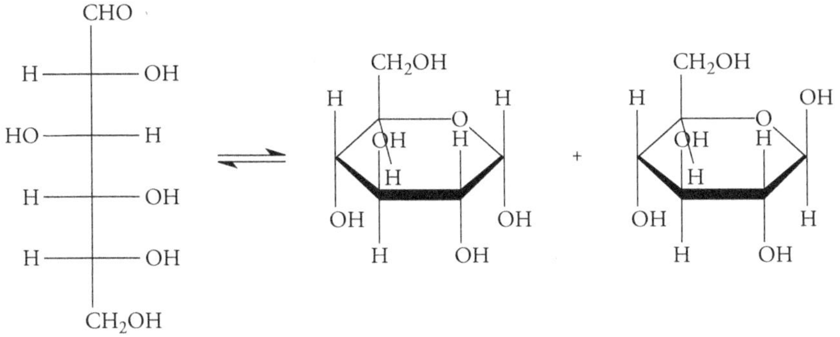

The ring forms will not give a positive Benedict's test. In one solution, 43% of the glucose is in the "straight" chain form. What percent of the glucose in this solution will react to give a positive Benedict's test?
(A) 0%
(B) 43%
(C) 57%
(D) 100%

428. This is the structure of adenine:

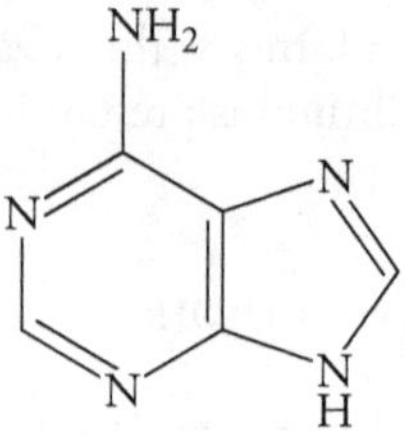

Which of the following is an acceptable resonance form for adenine?

(A)

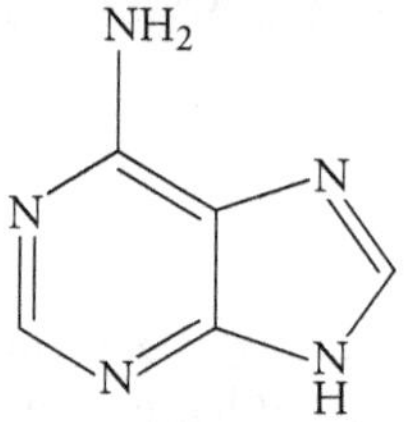

(B)

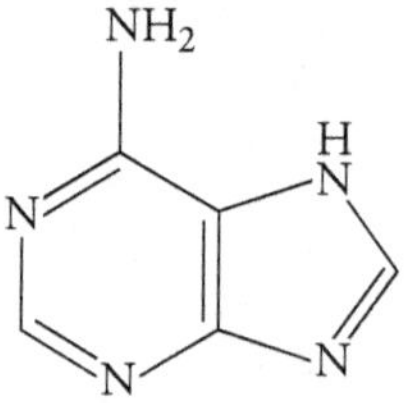

(C)

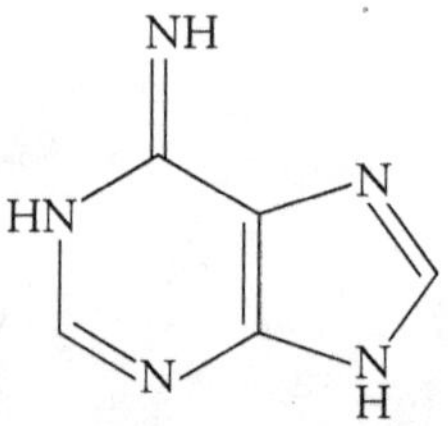

(D)

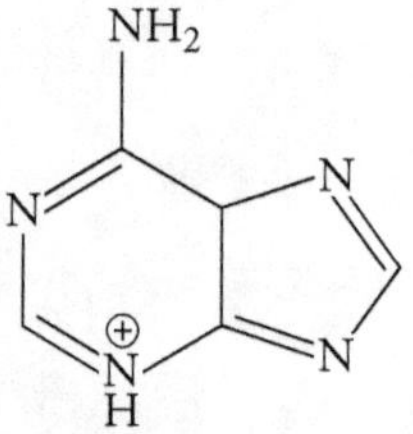

429. The drug thalidomide was introduced in the late 1950s to alleviate the morning sickness that often accompanies pregnancy. Synthesis of the drug resulted in a racemic mixture. One of the enantiomers was the active ingredient for treating morning sickness, while the other enantiomer led to birth defects. The following is one of the enantiomers of thalidomide. What are the expected hybridizations of the two nitrogen atoms in thalidomide?

(A) N-1 is sp^3, and N-2 is sp^2.
(B) N-1 is sp, and N-2 is sp^2.
(C) N-1 is sp^2, and N-2 is sp^3.
(D) N-1 is sp^3, and N-2 is sp^3.

CHAPTER 8

Final Review

430. How are the compounds cis-1,2-dibromopentene and trans-1,2-dibromopentene related?
    (A) They are geometric isomers.
    (B) They are diastereomers.
    (C) They are conformers.
    (D) They are structural isomers.

431. Which of the following is an acid anhydride?
    (A) [structure: CH₃CH₂—O—H]
    (B) [structure: CH₃—O—CH₂CH₃]
    (C) [structure: CH₃C(=O)—O—C(=O)CH₃]
    (D) [structure: CH₃C(=O)—O—CH₃]

432. Chymostatin is a drug used to interfere with the action of the enzyme chymotrypsin. The drug preferentially binds to the active site of chymotrypsin to reduce its activity. What type of interaction is this?
    (A) Competitive inhibition
    (B) Noncompetitive inhibition
    (C) Negative feedback
    (D) Nucleophilic attack

433. What is the product of the reaction of CD_3OH with isopropylmagnesium bromide?

(A) 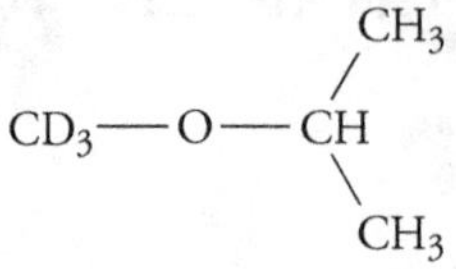 $CD_3-O-CH(CH_3)CH_3$

(B) 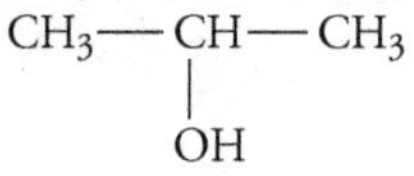 $CH_3-CH(OH)-CH_3$

(C) 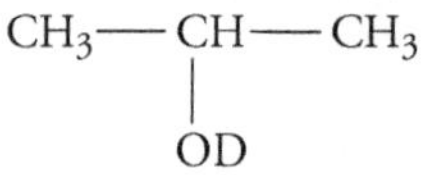 $CH_3-CH(OD)-CH_3$

(D) $CH_3-CH_2-CH_3$

434. It is possible to oxidize both $CH_3CH_2CHOHCH_2CH_3$ and $CH_3CH_2CDOHCH_2CH_3$ with acidic $K_2Cr_2O_7$. How do the two rates of reaction compare?

(A) The deuterated form reacts more slowly.
(B) The deuterated form reacts faster.
(C) The two forms react at the same rate.
(D) It is not possible to predict the relative rates.

435. Which of the following compounds will extract into sodium bicarbonate, but not into ether or dilute hydrochloric acid?

(A)  pentyl-NH_2

(B) 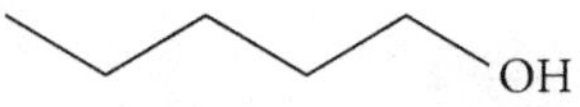 pentyl-OH

(C) 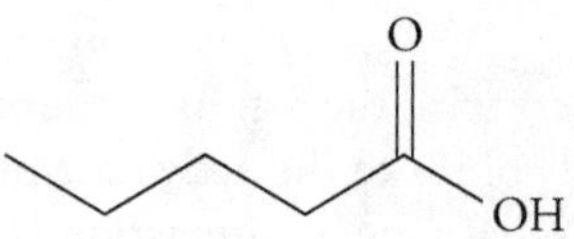 butanoic acid (—COOH)

(D) 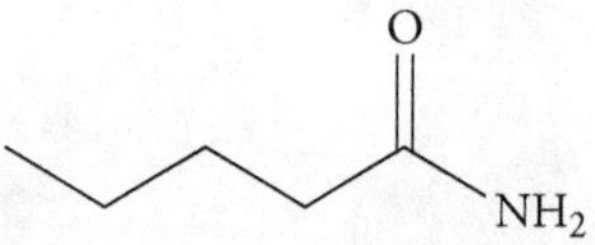 butanamide (—CONH$_2$)

436. What is the major product of the reaction of (R)-2-bromopentane with excess hydroxide ion?
    (A) (S)-2-pentanol
    (B) (R)-2-pentanol
    (C) Racemic 2-pentanol
    (D) 2-bromo-1-pentanol

437. Which of the following compounds would undergo the fastest dehydration when treated with concentrated phosphoric acid?

    (A)  OH

    (B) structure with OH

    (C) structure with OH

    (D) structure with OH

438. A polymer has the following structure:

    $(\text{-CH}_2\text{-CCl=CCl-CH}_2\text{-})_n$

    What is the structure of the monomer?
    (A) $CH_2=CHCl\text{-}CHCl=CH_2$
    (B) $CH_3CCl=CCl\text{-}CH_3$
    (C) $CH_2=CHCl$
    (D) $CH_2=CHCl$ plus $CH_2=CH_2$

439. Which of the following is NOT true concerning cyclohexyl lithium?
    (A) It is a strong oxidizing agent.
    (B) It reacts with water to form cyclohexane.
    (C) It is a weak reducing agent.
    (D) It is formed when cyclohexyl bromide reacts with lithium metal.

440. A scientist wishes to prepare the following molecule by reaction with bromine:

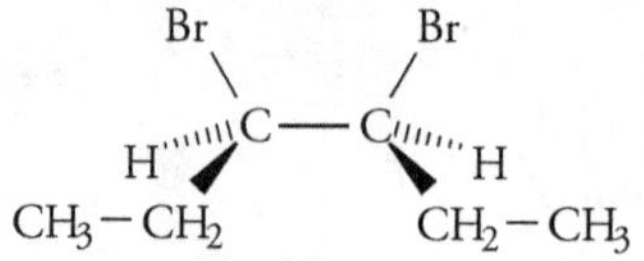

$$\underset{CH_3-CH_2}{\overset{Br}{\underset{H}{\diagdown}}}C-C\underset{CH_2-CH_3}{\overset{Br}{\underset{H}{\diagup}}}$$

What is the other reactant?
(A) trans-3-hexene
(B) cis-3-hexene
(C) 2,3-hexadiene
(D) 2-hexene

441. Which of the following is a correct Lewis structure for one of the resonance forms of the bicarbonate ion?

(A)

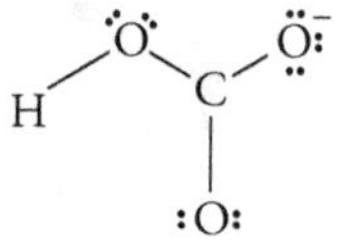

(B)

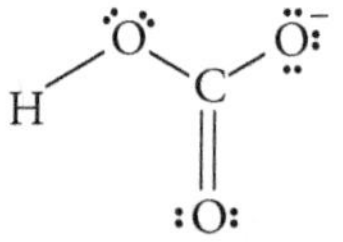

(C)

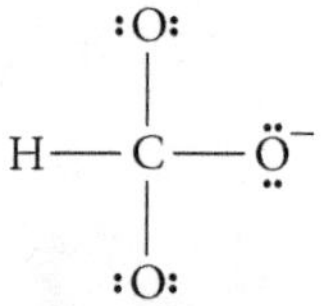

(D)

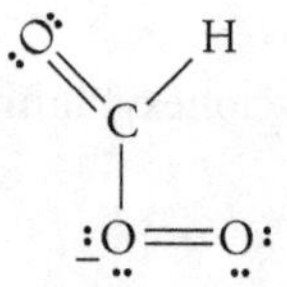

442. Rank the following resonance structures in order of importance from lowest contributor to highest contributor.

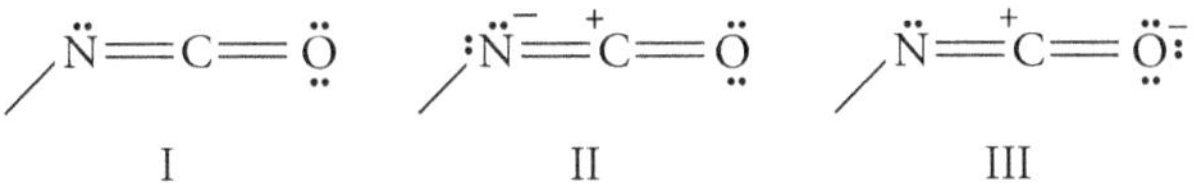

$$\text{N}=\text{C}=\ddot{\text{O}} \quad \text{:}\ddot{\text{N}}^-=\overset{+}{\text{C}}=\ddot{\text{O}} \quad \ddot{\text{N}}=\overset{+}{\text{C}}=\ddot{\text{O}}\text{:}^-$$
$$\text{I} \qquad\qquad \text{II} \qquad\qquad \text{III}$$

(A) II < III < I
(B) I < III < II
(C) II < I < III
(D) III < II < I

443. A researcher isolates the product of a reaction and determines it to have a melting point of 112°C. Later, she learns that there was a very small amount of an impurity present with a melting point of 145°C. How does the 112°C melting point compare to the melting point of the pure compound?

(A) It is lower.
(B) It is higher.
(C) It is the same.
(D) It is impossible to determine without knowing the identity of the impurity.

444. Which of the following describes a two-step reaction where the first step is rate-determining and the overall process is exothermic?

(A)
Reaction coordinate

(B) E
Reaction coordinate

(C)
Reaction coordinate

(D) E
Reaction coordinate

445. Compound II is in equilibrium with Compounds I and III. The following energy profile illustrates this relationship.

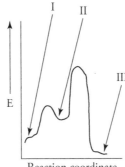

Reaction coordinate

Which of the following statements is true?

(A) I forms more rapidly than III; however, III is more stable.
(B) I forms more rapidly than III, and I is more stable.
(C) III forms more rapidly than I; however, I is more stable.
(D) III forms more rapidly than I, and I is more stable.

446. Which of the following is the best representation of the first step in the substitution reaction of 2-bromobutane with cyanide ion?

(A)

(B)

(C)

(D)

447. What are the hybridization and bond angles of the indicated atom?

(A) sp² and 120°
(B) sp³ and 109.5°
(C) sp² and 90°
(D) sp³ and 120°

448. Which of the following compounds has the highest boiling point?

(A) [structure]

(B) [structure with OH]

(C) [structure with NH₂]

(D) [structure with C=O and OH]

449. Which of the following is the most soluble in cyclohexane?

(A) [cyclohexene with COOH]

(B) [structure with OH]

(C) [branched structure]

(D) [structure with C=O]

450. Which of the following is the strongest acid?
    (A) CH₃CH₂CH₂CH₂OH
    (B) CH₃CH₂NHCH₂CH₃
    (C) CH₃CH₂CH₂CH₂NH₂
    (D) CH₃CH₂CH₂CH₂CH₃

451. Which of the following compounds yields only one monochloro product upon free radical chlorination?

(A)

(B)

(C)

(D)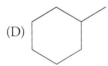

452. An organic chemist performs a free radical bromination of 3-methylpentane. Which of the following is the major product?

(A)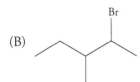

(B)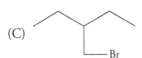

(C)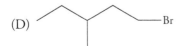

(D)

453. Many biologically important compounds, such as proteins, are amides, while others, such as fats, are esters. In the laboratory, the most efficient methods of producing amides and esters use acid chlorides or anhydrides. However, cells use alternative methods. Why is this the case?
    (A) Acid chlorides and anhydrides are too reactive to occur in cells.
    (B) Acid chlorides and anhydrides cannot cross the lipid bilayer and enter the cell.
    (C) Acid chlorides and anhydrides give different products under physiological conditions.
    (D) Acid chlorides and anhydrides do not form under physiological conditions.

454. What is the hybridization of atoms I and II in the following compound?

    (A) I is sp^3, and II is sp^2.
    (B) I is sp^2, and II is sp^2.
    (C) I is sp^2, and II is sp^3.
    (D) I is sp^3, and II is sp^3.

455. Aromatic compounds react with sodium dissolved in liquid ammonia to give two possible products through the Birch reduction. The reaction and possible products are as follows:

    Product I forms if R is electron-donating, and Product II forms if R is electron-withdrawing. Why is there a preference for Product I or II?
    (A) Resonance stabilization
    (B) Induced electrophilic substitution
    (C) Induced nucleophilic substitution
    (D) Ease of oxidation

456. The synthesis of many addition polymers begins with an initiator such as benzoyl peroxide. This is the structure of benzoyl peroxide:

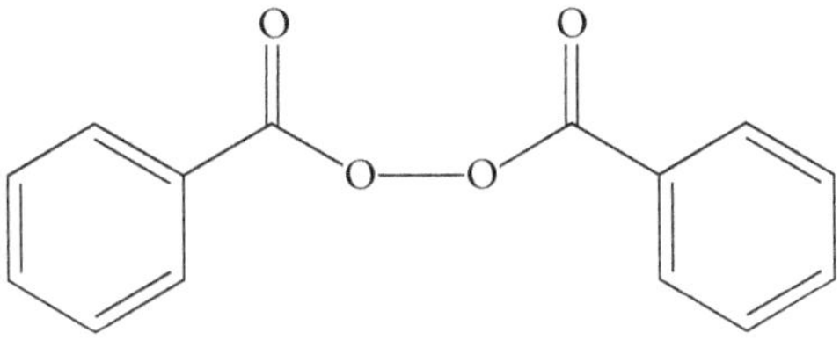

The reaction begins with the homolytic cleavage of the O–O bond to form two benzoyl radicals. The benzoyl radical can attack a substituted alkene such as styrene to begin a chain reaction. The attack of the alkene by the benzoyl radical is:

(A) A propagation step.
(B) An initiation step.
(C) A termination step.
(D) A complexation step.

457. The Kolbe electrolytic synthesis produces hydrocarbons from carboxylic acids. The general reaction is:

$$2 \text{ RCOO}^- \xrightarrow{\text{electrolysis}} \text{R–R} + 2 \text{ CO}_2$$

What type of reaction is this?
(A) A reduction
(B) An oxidation
(C) A neutralization
(D) A free radical addition

458. The Wacker oxidation can convert terminal alkenes to methyl ketones by the reaction of molecular oxygen. The catalyst is a mixture of palladium and copper chloride. The reaction is:

$$R-CH=CH_2 \xrightarrow{O_2, PdCl_2/CuCl_2} R-C(=O)-CH_3$$

What change in the IR spectrum could be used to monitor the progress of the reaction?

(A) The disappearance of a band near 1650 cm^{-1} and the increase in a band near 1715 cm^{-1}
(B) The disappearance of a band near 1650 cm^{-1} and the increase in a band near 3620 cm^{-1}
(C) The disappearance of a band near 2240 cm^{-1} and the increase in a band near 1715 cm^{-1}
(D) The disappearance of a band near 2180 cm^{-1} and the increase in a band near 1715 cm^{-1}

459. Which of the following amino acids CANNOT form a coordinate covalent bond to a metal ion in a metalloenzyme?

(A) H$_2$N—CH(CH$_2$SH)—C(=O)—OH

(B) H$_2$N—CH((CH$_2$)$_4$NH$_2$)—C(=O)—OH

(C) H$_2$N—CH(CH$_2$COOH)—C(=O)—OH

(D) H$_2$N—CH(CH(CH$_3$)CH$_3$)—C(=O)—OH

460. Urea, a metabolic waste product, exists in the following resonance forms:

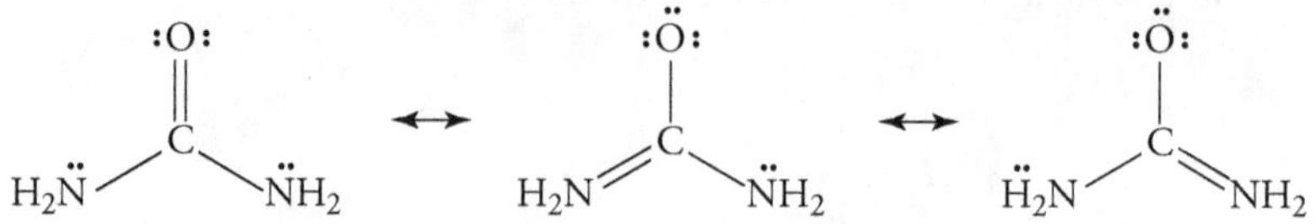

Considering all the resonance forms, what are the charges on the oxygen atom and the nitrogen atoms?
(A) O is δ–, and both Ns are δ+.
(B) O is δ+, and both Ns are δ–.
(C) O is δ–, and both Ns are 0.
(D) All atoms have 0 charge.

461. Urea, a metabolic waste product, exists in the following resonance forms:

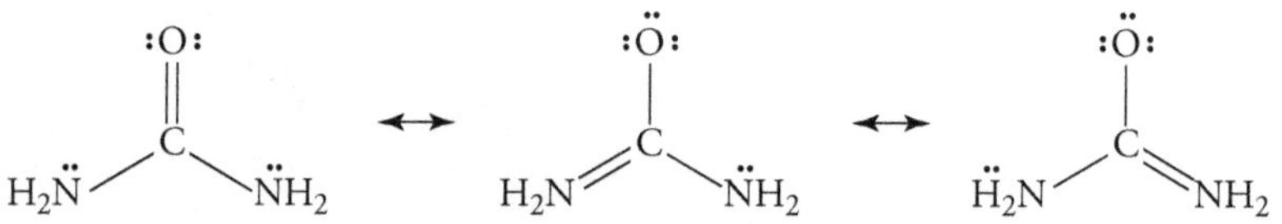

What is the shape of a urea molecule?
(A) Trigonal planar
(B) Tetrahedral
(C) Trigonal pyramidal
(D) T-shaped

462. The first step in the preparation of TNT (trinitrotoluene) is the mononitration of toluene. (The toluene undergoes two additional nitrations to yield the final product.) The first step in the formation of TNT is:

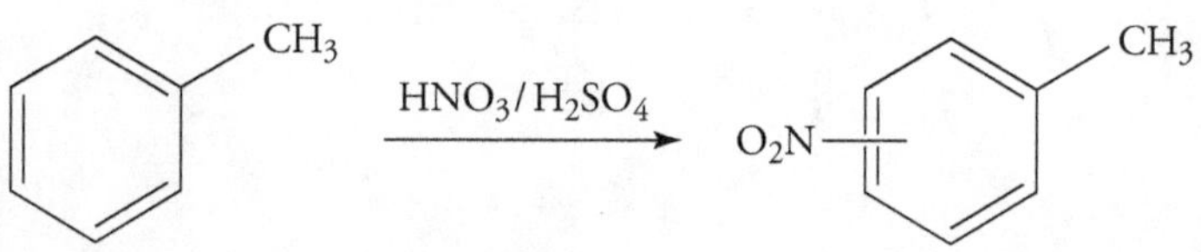

What species actually attacks the aromatic ring in this step?
(A) NO_2^+
(B) NO_3^-
(C) H^+
(D) HSO_4^-

463. The first step in the preparation of TNT (trinitrotoluene) is the mononitration of toluene. (The toluene undergoes two additional nitrations to yield the final product.) The first step in the formation of TNT is:

$$\text{C}_6\text{H}_5\text{CH}_3 \xrightarrow{\text{HNO}_3/\text{H}_2\text{SO}_4} \text{O}_2\text{N-C}_6\text{H}_4\text{-CH}_3$$

What is the most likely position for the NO_2 relative to the CH_3?
(A) Ortho or para
(B) Ortho or meta
(C) Meta or para
(D) Meta only

464. The melting points for the nitrobenzoic acids are in the following table:

Ortho-nitrobenzoic acid	147°C
Meta-nitrobenzoic acid	140°C
Para-nitrobenzoic acid	242°C

Why is the melting point of para-nitrobenzoic acid so much higher than that of either of the other two compounds in the table?
(A) Para-nitrobenzoic acid has intermolecular hydrogen bonding, and the other two have intramolecular hydrogen bonding.
(B) Para-nitrobenzoic acid has intramolecular hydrogen bonding, and the other two have intermolecular hydrogen bonding.
(C) Only para-nitrobenzoic acid can participate in hydrogen bonding.
(D) Para-nitrobenzoic acid undergoes an acid-base reaction with the neighboring molecules to form a covalent bond between the molecules.

465. The mononitration of phenol yields a mixture of ortho-nitrophenol and para-nitrophenol. Ideally, what will be the melting point of a mixture containing equal amounts of these two isomers?
(A) Below the melting point of the lower-melting compound
(B) Above the melting point of the higher-melting compound
(C) The average of the two melting points
(D) Slightly above the average of the two melting points

466. The nitration of benzene is an electrophilic substitution reaction in which a nitro group replaces a hydrogen atom. During this reaction, what are the initial, intermediate, and final hybridizations of the carbon atom that is attacked by the nitro group?
 (A) sp^2, sp^3, sp^2
 (B) sp^2, sp^2, sp^2
 (C) sp^2, sp^2, sp^3
 (D) sp^3, sp^2, sp^2

467. The presence of a free radical is necessary for a free radical mechanism. Which of the following is a free radical?
 (A) NO_2
 (B) H^-
 (C) Cl^-
 (D) Cl_2

468. The bond angles in the methyl radical are all 120°. What type of orbital contains the unpaired electron?
 (A) p
 (B) sp^2
 (C) sp^3
 (D) s

469. Every amino acid has at least two pK_a values, one for the carboxylic acid group and one for the amino group. Some amino acids have an additional pK_a value for the side chain. For aspartic acid, the pK_a values are 2.1 and 9.8 (carboxylic acid and amino groups, respectively). The side chain has a pK_a of 3.9. What is the predominant structure for aspartic acid at pH 7?

(A)
$$H_2N-CH(CH_2-C(=O)-OH)-C(=O)-OH$$

(B)
$$H_3\overset{+}{N}-CH(CH_2-C(=O)-O^{-})-C(=O)-OH$$

(C)
$$H_3\overset{+}{N}-CH(CH_2-C(=O)-O^{-})-C(=O)-O^{-}$$

(D)
$$H_3\overset{+}{N}-CH(CH_2-C(=O)-OH)-C(=O)-O^{-}$$

470. Amino acids contain an amine group and a carboxylic acid group. For this reason, amino acids are classified as amphoteric substances. Why are amino acids classified as amphoteric substances?

(A) They may behave as either a Lewis acid or an Arrhenius acid.
(B) They exist as enantiomeric pairs.
(C) They exist as pairs of diastereomers.
(D) They may behave as either acids or bases.

471. In the vapor phase, chloroacetic acid exists as dimers. A dimer is a pair of strongly interacting molecules. Which of the following is the most likely structure of this dimer?

472. Every amino acid has at least two pK_a values, one for the carboxylic acid group and one for the amino group. Some amino acids have an additional pK_a value for the side chain. Which of the following will have the lowest pK_a for its carboxylic acid group?

(A)
$$H_2N-CH(-CH_2-CH_2-C(=O)-OH)-C(=O)-OH$$

(B)
$$H_2N-CH(-CH_2-CH(CH_3)-CH_3)-C(=O)-OH$$

(C)
$$H_2N-CH(-CH(CH_3)-CH_3)-C(=O)-OH$$

(D) $H_2N-CH(-CH_3)-C(=O)-OH$

473. As you saw earlier, the following carbohydrate sequence shows the terminal sequence of the Type A determinant protein in blood:

Which of the subunits can change its stereochemistry by mutarotation?
(A) Unit A
(B) Unit B
(C) Unit D
(D) None

474. Condensation polymers form when monomers join through condensation reactions. In a condensation reaction, two groups join, with the expulsion of a small molecule such as water. Polymers such as polyesters, nylons, and proteins are examples of condensation polymers. Which of the following is LEAST likely to serve as a monomer for the formation of a condensation polymer?
(A) $CH_3CH_2CH_2OH$
(B) $HOCH_2CH_2COCl$
(C) $H_2NCH_2CH_2C(O)OC(O)CH_2CH_2NH_2$
(D) $HOCH_2CH_2CH_2COOH$

475. Which is the most stable conformer of 1-t-butyl-4-propyl cyclohexane?
(A) Both are equatorial.
(B) The t-butyl is equatorial, and the propyl is axial.
(C) The t-butyl is axial, and the propyl is equatorial.
(D) Both are axial.

476. Which amino acid side chain may undergo hydrolysis during hydrolysis of a protein?
   (A) Valine
   (B) Phenylalanine
   (C) Leucine
   (D) Asparagine

477. Nylon-66 is a polymer formed from 1,6-diaminohexane and adipic acid (1,6-hexanedioic acid). What is the classification of the polymer formed?
   (A) Polyamide
   (B) Polyester
   (C) Polycarboxylate
   (D) Polysaccharide

478. What is the product of the free radical monobromination of (R)-3-ethyl-2,2-dimethylhexane?
   (A) The (S) product
   (B) The (R) product
   (C) A racemic mixture
   (D) The meso product

479. What is the maximum number of products isolated from the following reaction?

   (A) 2
   (B) 5
   (C) 7
   (D) 3

480. How many chiral centers are in the following compound?

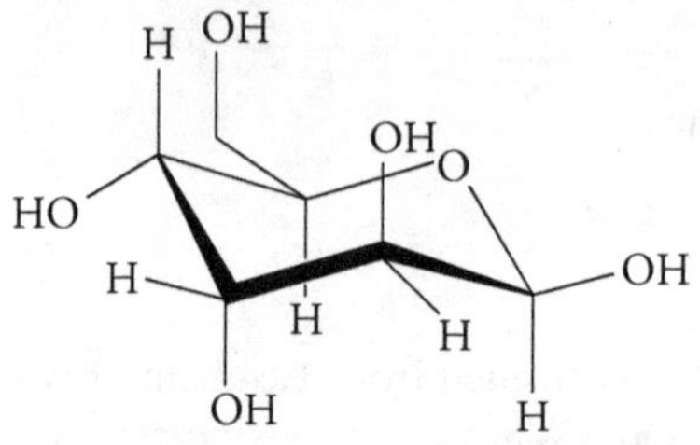

(A) 5
(B) 7
(C) 3
(D) 4

481. How many σ and π bonds are in the following compound?

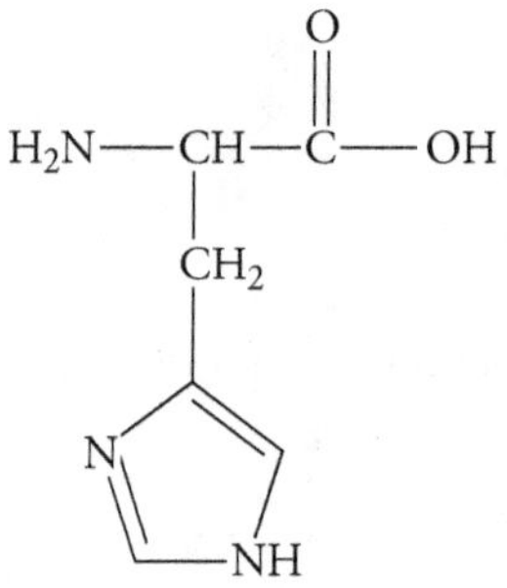

(A) 27 σ and 2 π
(B) 19 σ and 2 π
(C) 17 σ and 3 π
(D) 20 σ and 3 π

482. What is the degree of unsaturation in the following molecule?

$$H_2N-CH(CH_2-\text{imidazole})-C(=O)-OH$$

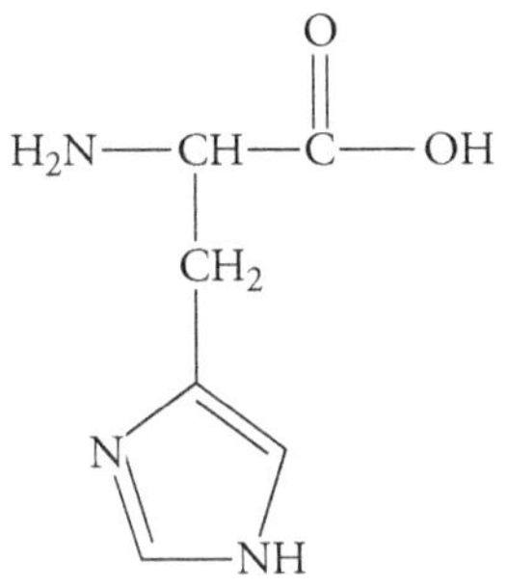

(A) 6
(B) 5
(C) 4
(D) 3

483. How many chiral centers are in the following molecule?

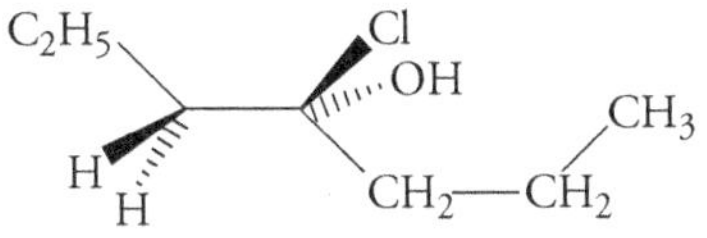

(A) 0
(B) 2
(C) 4
(D) 5

484. Which is the correct order of increasing acidity?
(A) sp-hybridized CH bonds < alcohols < phenols < carboxylic acids < sulfonic acids
(B) sp-hybridized CH bonds > alcohols > phenols > carboxylic acids > sulfonic acids
(C) sp-hybridized CH bonds < alcohols ≈ phenols < carboxylic acids ≈ sulfonic acids
(D) sp-hybridized CH bonds ≈ alcohols < phenols ≈ carboxylic acids < sulfonic acids

485. Which of the following is the most stable conformer of butane?

(A) 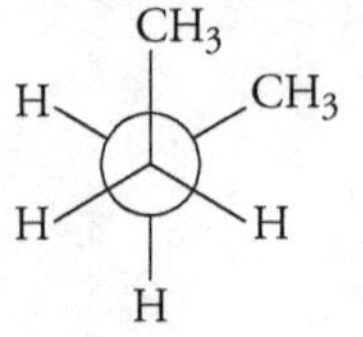 [Newman projection: front CH3 up-right, H upper-left, H lower-left, H bottom; back CH3 upper-right, H right, H bottom]

(B) 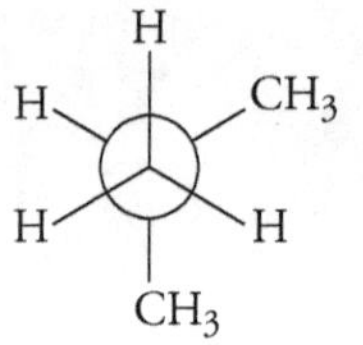 [Newman projection: front H up, H upper-left, H lower-left, CH3 bottom; back CH3 upper-right, H right, H bottom]

(C) 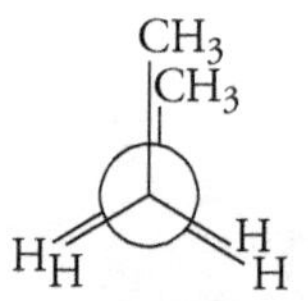 [Newman projection: front CH3 up; back CH3 up, H and H lower positions]

(D) 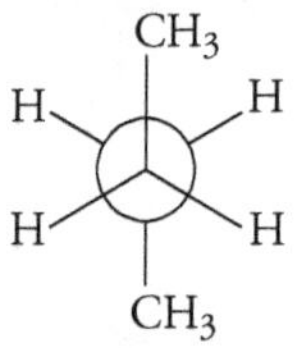 [Newman projection: front CH3 up, H upper-left, H lower-left, CH3 bottom; back H upper-right, H right, H bottom]

486. How many monochlorinated isomers (including cis and trans forms) will the following reaction produce?

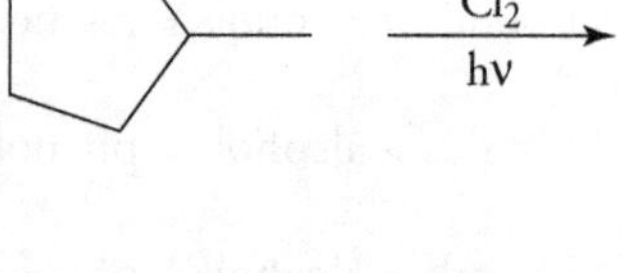

$$\text{cyclopentyl-CH}_3 \xrightarrow[\text{hv}]{Cl_2}$$

(A) 8
(B) 6
(C) 4
(D) 3

487. Which of the following free radicals is the most stable?

(A)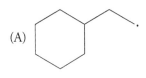

(B)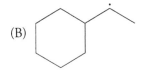

(C)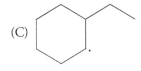

(D)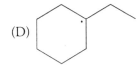

488. It requires a higher temperature to sulfonate benzoic acid than to sulfonate toluene. Why is this true?
   (A) The carboxylic acid group is deactivating, and the alkyl group is activating.
   (B) The carboxylic acid group is more deactivating than the weakly deactivating alkyl group.
   (C) The alkyl group is deactivating, and the carboxylic acid group is activating.
   (D) The alkyl group leads to less steric hindrance than the carboxylic acid group.

489. Which of the following is the strongest acid?
   (A) HCl
   (B) H_2O
   (C) HBr
   (D) NH_3

490. The complete combustion of 1 mole of octane produces:
   (A) 8 moles of CO_2 and 8 moles of H_2O.
   (B) 16 moles of CO_2 and 18 moles of H_2O.
   (C) 8 moles of CO_2 and 10 moles of H_2O.
   (D) 8 moles of CO_2 and 9 moles of H_2O.

491. The free radical monochlorination of methylpropane may produce how many different products?
(A) 2
(B) 1
(C) 3
(D) 4

492. One experiment in the investigation of the reaction of ethyl iodide with hydroxide ion found that tripling the concentration of both the ethyl iodide and the hydroxide ion caused the rate to:
(A) Remain the same.
(B) Increase × 3.
(C) Increase × 9.
(D) Increase × 6.

493. The first step in a hydroboration reaction is an attack on an alkene by BH_3. In this case, BH_3 is behaving as:
(A) A Lewis acid.
(B) A Lewis base.
(C) A nucleophile.
(D) An Arrhenius acid.

494. Which of the following molecules would NOT be a monomer in a polymerization reaction?
(A) $H_2NCH_2CH_2CH_2NH_2$
(B) $HOCH_2CH_2CH_2OH$
(C) $CH_3CH_2CH_2OH$
(D) $HOOC(CH_2)_4COOH$

495. Which of the following is the strongest base?
(A) CH_3^-
(B) OH^-
(C) NH_2^-
(D) OCH_3^-

496. For which of the following molecules is hydrogen bonding an important factor influencing the physical properties?
(A) HBr
(B) CH_2F_2
(C) CH_3CH_3
(D) $NH(CH_3)_2$

497. What is the relationship between the following two molecules?

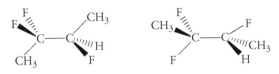

(A) They are enantiomers.
(B) They are the same.
(C) They are diastereomers.
(D) They are meso.

498. Why is trifluoroacetic acid a stronger acid than trichloroacetic acid?
(A) Fluorine is more electronegative than chlorine; therefore, there is a greater inductive effect.
(B) Chlorine is more electronegative than fluorine; therefore, there is a greater inductive effect.
(C) Fluorine can participate in hydrogen bonding, and chlorine cannot.
(D) Hydrolysis forms hydrofluoric acid, which is a stronger acid than hydrochloric acid.

499. The following equilibria apply to all amino acids:

$$H_3\overset{\oplus}{N}-CH(R)-C(=O)-OH \underset{+H^+}{\overset{-H^+}{\rightleftharpoons}} H_3\overset{\oplus}{N}-CH(R)-C(=O)-O^{\ominus} \underset{+H^+}{\overset{-H^+}{\rightleftharpoons}} H_2N-CH(R)-C(=O)-O^{\ominus}$$

The central form is the zwitterion form. Which of the following statements is true?
(A) The zwitterion is in equilibrium with the other forms, so it is neither an acid nor a base.
(B) The zwitterion can act either as an acid or as a base.
(C) All three forms can act as a base.
(D) All three forms can act as an acid.

500. It is possible to produce mostly one alkyl halide via the following reaction:

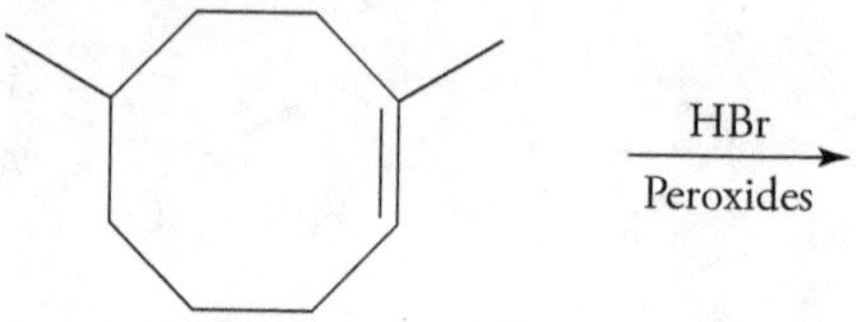

[structure: 1,3-dimethylcyclooctene] $\xrightarrow[\text{Peroxides}]{\text{HBr}}$

Which of the following is the major product of this reaction?

(A) 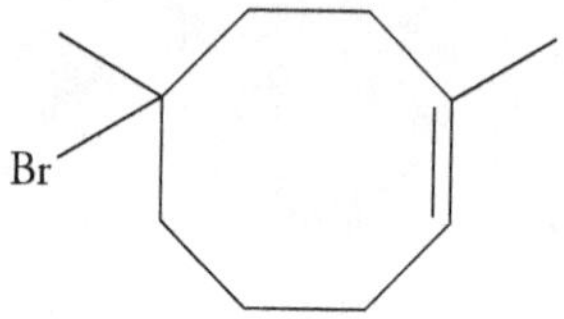 [structure with Br]

(B) 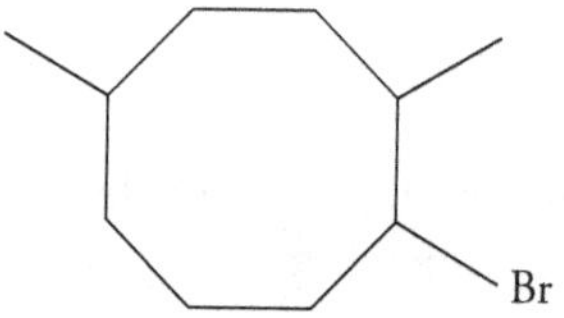 [structure with Br]

(C) 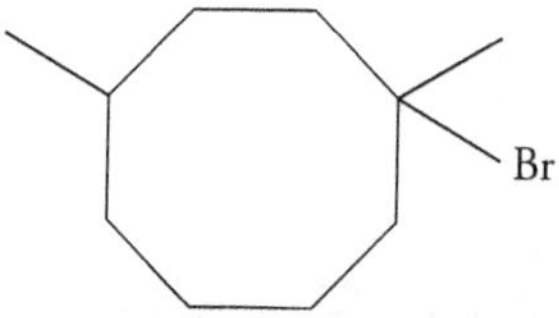 [structure with Br]

(D) 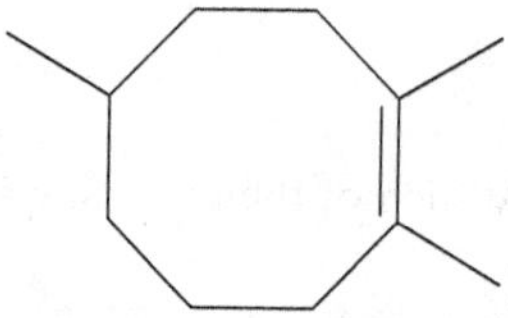 [structure with Br]

One more. A biochemist plans on synthesizing a small polypeptide. The synthesis is a three-step process. The experimental percent yields of each of the steps are 90%, 30%, and 70%, respectively. What is the overall percent yield?
(A) 60%
(B) 90%
(C) 30%
(D) 20%

ANSWERS

Chapter 1: The Fundamentals

1. **(B)** Single and double bonds hold all the carbon atoms in this nanotube together; therefore, the answer is sp^2 hybridization.

2. **(B)** The nitrogen atoms are triply bonded to each other. A triple bond involves sp hybridization.

3. **(D)** Carbon 1 is triply bonded (sp hybridization). Carbon 2 has a single bond to another carbon, two single bonds to hydrogen atoms, and one bond to a phosphorus atom. Carbon 3 has a double bond to an oxygen atom and two single bonds, one to another carbon atom and one to a nitrogen atom.

4. **(C)** Beginning at the triple-bonded end, the bond is between two sp hybridized carbon atoms (carbons 1 and 2), between an sp and an sp^3 hybridized carbon (carbons 2 and 3), and between two sp^3 hybridized carbons (carbons 3 and 4); therefore, sp and sp^3 hybridized orbitals are utilized.

5. **(D)** There are five on the ring containing the nitrogen, one on the methyl attached to the nitrogen, and two in the ethyl group, for a total of eight.

6. **(A)** All the nitrogen atoms indicated have a lone pair of electrons on the nitrogen. Nitrogen I is triple bonded and has a lone pair, indicating sp hybridization; nitrogen II has three single bonds and a lone pair, indicating sp^3 hybridization; and nitrogen III has a double bond, a single bond, and a lone pair, indicating sp^2 hybridization.

7. **(A)** The methyl cation has three groups around it, corresponding to sp^2 hybridization, leaving an unhybridized p orbital to interact. The methyl anion would have four groups, three bonding pairs, and one nonbonding pair, corresponding to sp^3 hybridization.

8. **(A)** You can eliminate answers B and D because of the charge separation, and answer A has the lowest formal charges on the atoms. Answer C features 10 electrons around carbon.

9. **(D)** Structure III is an ionic compound, not a resonance form. Therefore, that eliminates answers A, B, and C.

10. **(B)** The presence of the highly electronegative fluorine atoms withdraws electron density from the –O-H bond by an induction effect, weakening the bond and therefore increasing the acidity. The more fluorine atoms there are or the closer the fluorine atoms are to the –OH, the greater the inductive effect and the stronger the acid.

11. **(D)** In ortho-nitrophenol there is resonance stabilization of the anion formed. There is no comparable stabilization in the case of phenol.

12. (A) Answer A is not a cyclohexane but a cyclohexene.

13. (D) The conjugate base generated by the loss of the hydrogen ion through resonance increases the negative charge of the aromatic system. An electron-withdrawing group will stabilize this anion; however, an electron-donating group would have the opposite effect and make the acid weaker. The amino group is an electron-donating group. The amino group does not destabilize the aromatic system, and it is too small to create any significant steric hindrance.

14. (B) The triflate anion has the ability to exist in resonance forms with delocalization of the negative charge. This resonance stabilizes the structure, making it more stable than the hydroxide ion and a better leaving group.

15. (B) Here you have a six-carbon backbone with alternating single and double bonds. The bond between C(4) and C(5) is a single bond. Since there are sp^2 bonds on either side of it, this bond has more s character than the sp^3 hybrid bond in a normal alkane, and therefore it is shorter.

16. (A) The lone pair on the nitrogen resonates to form a double bond to the carbon, with one of the pairs from the double bond moving to the oxygen atom to give a stabilizing resonance form.

17. (A) The base is a catalyst in the hydrolysis of an ester because changing its concentration has no effect on the identity of the products of the reaction, and the base is not consumed in the reaction. The conversion to the carboxylate ion is a distracter.

18. (D) Bonds 1 and 2 should be of equal length, since resonance will occur and they will be somewhere between a single and a double bond in length. The resonance makes bonds 1 and 2 equal in length. Bond 3 is a double bond (without significant resonance) and will be shorter than bonds 1 and 2.

19. (A) The hydrogen labeled 3 will have the weakest bond (making it the most acidic) because of the two carbonyl oxygen withdrawing electron density. In addition, the loss of an H^+ from position 3 leaves a resonance-stabilized carbanion. The presence of the two adjacent carbonyls is responsible for the resonance stabilization. The H at position 2 is more acidic than the H at position 1 than the H at position 4 because of the high electronegativity of the fluorine. The adjacent carbonyl, because of resonance, makes the hydrogen at position 2 more acidic than the hydrogen at position 1.

20. (B) Do not forget the 9 implied hydrogen atoms for 9 σ bonds. Each other bond, whether single, double, or triple, contains a σ bond. The total is 32 σ bonds.

21. (C) Amines have a nitrogen atom attached to hydrogen atoms or R groups. There are two nitrogen atoms present: one a nitrogen-carbon triple bond (nitrile), and the other a nitrogen attached to two hydrogen atoms and a carbonyl (amide). There is also a phosphate present.

22. (A) In this compound, the alcohol is the more important functional group; therefore, the numbering should give it the lowest number (and the final position in the name) versus the alkyne. There are six carbons present with a triple bond, and numbering from right to left (centering on the more important alcohol), the alcohol group is at carbon 3, making it a hexyn-3-ol. Since the triple bond is between carbons 5 and 6, it would be a 5-hexyn-3-ol. There are no cis/trans options.

23. (D) Amphoteric means that it could behave either as an acid or as a base. The alcohol group is not the key in aqueous solution. Terminal alkynes are acidic.

24. (B) Naphthalene is two fused benzene rings with the chemical formula $C_{10}H_8$. There are 11 carbon-to-carbon σ bonds and 8 carbon-to-hydrogen σ bonds (single bonds). In addition, there are five π bonds in the ring system (one per double bond). Resonance does not alter the total.

25. (A) For the compound to be saturated, there would need to be 22 hydrogen atoms, not 14. The degree of unsaturation would be $(22 - 14)/2 = 4$, so there would be a total of four π bonds and rings. Hydrogenation attacks π bonds; therefore, there are three π bonds. Of the possible answers, the only reasonable one is three π bonds, leaving one ring.

26. (C) The formula $C_{10}H_{23}NO$ is saturated; therefore, it cannot belong to any compound containing a double bond. The amide contains a carbonyl group, which has a carbon-oxygen double bond.

27. (D) The C-O bond in methanol is a single bond. There are C-O double bonds in carbon dioxide and a triple bond in carbon monoxide. The carbonate ion is a resonating system containing one double and two single bonds to give an average of 1.33. The more bonds there are, the shorter the separation is. The order is $3 < 2 < 1.33 < 1$.

28. (B) In general, the most stable conformer is the anti; therefore, eliminate all choices beginning with anti. The least stable conformer is eclipsed; therefore, the answer is B.

29. (B) The compound is a heptene (seven-carbon backbone, with a double bond). Numbering to give the double bond the lowest possible number makes it a 3-heptene, and having the R groups on the same side of the double bond makes it a (Z).

30. (D) There are three chiral carbon atoms. These are carbon atoms 2, 4, and 5. There will be 2^n stereoisomers, with $n = 3$; therefore, there are 8 stereoisomers.

31. (A) The longest chain has six carbon atoms and only single bonds, which makes it a hexane. The two substituents (bromo and chloro) alphabetically give bromochlorohexane. Numbering from the end that gives the substituents the lowest numbers gives the answer.

32. (C) An acid anhydride consists of two carbonyl groups linked by an oxygen atom.

33. (C) The isomers are n-hexane, 2-methylpentane, 3-methylpentane, 2,2-dimethylbutane, and 2,3-dimethylbutane.

34. (D) An isobutyl group is a 2-methylpropyl.

35. (A) The nitrogen adjacent to a carbonyl is an amide. Two carbon atoms make this a substituted acetamide. The –OH group is para (number 4) on the phenyl ring. The ring is attached to the N, making it an N-.

36. (B) The chain splits at the oxygen. There are five carbon atoms to the left (*pent*) and three carbon atoms to the right (*prop*). An oxygen atom adjacent to a carbonyl can be an ester (*-oate* suffix). The acid side of the ester has the carbonyl group. The part of an ester derived from a five-carbon chain is a pentanoate, with the carbonyl carbon being 1. If the carbonyl carbon is 1, then the fluorine is on 4. Combining these gives a 4-fluoropentanoate. The propyl group, from the alcohol, remains and, according to the nomenclature rules, is the first part of the name.

37. (D) Every ring is one degree of unsaturation, and every π bond is one degree of unsaturation. Answer A gives 3, answer B gives 3, answer C gives 3, and answer D gives 4.

38. (A) Assuming that sulfur is more electronegative than carbon and hydrogen, the sulfurs in the disulfide bonds are –1 and in the reduced form are –2.

39. (D) The longest chain has seven carbon atoms and only single bonds; therefore, this is a heptane. The branches are methyl, bromo, and hydroxyl (alcohol). The alcohol has highest priority; therefore, this is an alcohol, with the –OH being closest to the end of the chain. This gives heptan-2-ol. Following through on the numbering will yield 4-bromo and 5-methyl. Adding these to the root name (alphabetically) gives the answer.

40. (D) An acetal is $R_2C(OR)_2$. The Rs may be the same or different.

41. (C) The compound has an eight-carbon backbone with a double bond (octene) and a carboxylic acid group (octenoic acid). Numbering starts at the end closest to the acid end (2-octenoic acid), and the R groups are oriented on opposite sides of the double bond, making it an (E) conformation.

42. (A) There are two basic structures, propene and cyclopropane. There are three cyclopropane isomers, (1,1), (cis-1,2), and (trans-1,2). There are seven propene isomers, (1,1), (2,3), (3,3), (cis-1,2), (trans-1,2), (cis-1,3), and (trans-1,3).

43. (D) If the compound were saturated, it would have 2(12) + 2 = 26 hydrogen atoms, instead of 14. The degree of unsaturation is (26 – 14)/2 = 6.

44. (A) The compound can be a bromocyclopropane or a 1-bromopropene (cis or trans). But it cannot be a cis- or trans-2-bromopropene, because the bromine will be attached to the carbon with the methyl group and there are only hydrogen atoms on the other carbon—therefore, no cis or trans isomers are possible.

45. (B) In general, the "largest" substituent on a cyclohexane ring should be equatorial because there is less steric hindrance in this position.

46. (C) These are structural isomers of each other.

47. (C) Answer D is a four-membered ring (90° bond angle) and would have more bond strain than the other compounds, which are cyclohexanes. Answers A and B are the chair forms with a great deal of steric interaction because the groups are close. This is not the case in answer C, which is in the chair form.

48. (A) HCl is a strong acid and will be stronger than any of the organic acids. II will be the next strongest because of considerable resonance stabilization of the anion (conjugate base). I is next, due to minor resonance stabilization of the anion. III is an alcohol with no resonance stabilization possible, so it is the weakest acid.

49. (C) Carbon 1 has three groups around it, giving a trigonal planar geometry (120°).

50. (C) The two compounds are not mirror images of each other, so they cannot be enantiomers. They are not non-mirror-image stereoisomers, so they are not diastereomers. They are not conformers, because that refers to the free rotation around a double bond. They are regioisomers, or positional isomers, differing in the position of the groups attached to the double-bonded carbon.

51. (B) In conjugated systems like this one, the C-C single bonds are commonly shorter than normal ones because of the partial overlap of the unhybridized p orbitals of the two adjacent carbons, which leads to resonance.

52. (C) The trans isomer would be the most stable (lower energy). The cis isomer is the least stable (steric hindrance with the groups attached to the double-bonded carbons); therefore, it is the higher-energy form and will release more energy upon hydrogenation.

53. (A) There are many systems for calculating the degree of unsaturation. In general, take twice the number of carbon atoms, add 2, subtract the number of hydrogen atoms, add the number of nitrogen atoms, and ignore the oxygen atoms. Finally, divide the result by 2. The degree of unsaturation would be $[2(6) + 2 - 10 + 2 + 0]/2 = 3$.

Chapter 2: Isomers and Physical Properties

54. (D) The nitrogen can undergo inversion; therefore, it is rapidly changing from one orientation to the opposite orientation

55. (C) To determine the absolute configuration about a carbon atom, note the lowest-priority group and assign priority to the other groups. Beginning with the highest-priority group (Cl in this case), count toward the next highest group and end with the lowest group. Counting to the right (clockwise) is R, and counting to the left (counterclockwise) is S. The configuration about each carbon atom is independent of the configuration about the other carbon atom. Doing this, both carbons turn out to be S.

56. (A) Achiral compounds are those that are superimposable on their mirror images. I and II cannot be superimposed, but III can. It is a meso compound, as the chlorine atoms are cis, creating a symmetry plane between them.

57. (D) You are told that both have a chiral center; there is no way to determine the relationship (+ or −) of either compound. Changing atoms (F for Cl) may or may not retain the + or −. There is no information given or predictable about the type (direction) of the rotation.

58. (D) The glycine is achiral; therefore, it is possible to produce pure glycine. The other natural amino acids occur as D and L enantiomers. Laboratory synthesis gives a mixture of equal amounts of the D and L enantiomers. The mixture of equal amounts of the D and L enantiomers is called a racemic mixture. It would not be optically active (eliminating answer A); *mesomeric* and *diastereometric* are not terms used to describe enantiomers.

59. (B) A chiral center has four different groups attached to it. In 2,3-hydroxypentane, there are two chiral centers, the carbon atoms with hydroxy groups. Each of those carbons has two different R groups, a hydrogen atom and a hydroxy group, attached to it.

60. (A) It is impossible to know what the optical rotation of the product will be. The mechanism does cause inversion of the configuration; however, this does not mean inversion of optical rotation.

61. (A) The presence of the lone pair on the nitrogen atom leads to the formation of the double bond to it (N=X), and the nitrogen is best able to accommodate the positive charge.

62. (D) Assign the priorities to each atom (Br > F > -CH$_2$-C- > CH$_3$). Then see if the Br →F → and so on rotation is to the right (R) or to the left (S). Answer D has the proper absolute configuration on each atom.

63. (B) These are enantiomers.

64. (D) The trans fluorine atoms cancel, while in all other cases the fluorine atoms do not cancel.

65. (A) The compounds are isomers; therefore, their molar masses are the same. Only ethyl methyl amine is capable of forming hydrogen bonding, while trimethyl amine cannot.

66. (A) To determine the absolute configuration about a carbon atom, note the lowest-priority group and assign priority to the other groups. Beginning with the highest-priority group, count toward the next highest group and end with the lowest group. Counting to the right (clockwise) is R, and counting to the left (counterclockwise) is S. The configuration about each carbon atom is independent of the configuration about the other carbon atom. The carbon with the two carbons attached is not a stereocenter. For one carbon, you would count to the left (S), and for the other, you would count to the right (R).

67. (C) Each racemic mixture would contain two enantiomers: A, B and C, D. The possible combinations of products would then be AC, AD, BC, and BD, for a total of four products.

68. (D) Diastereomers can be easily separated from each other, whereas none of the others can be separated easily, if at all.

69. (C) Compound I is nonpolar because all four groups are the same. That eliminates answers A and D. Both II and III are polar and have nonzero dipole moments.

70. (A) Carbonyls are very polar, ethers are slightly polar, and alcohols are somewhere in between those two.

71. (A) Answer B is a hydrocarbon, so it will not be very soluble in water at all. Answer D can form very weak dipole attractions, but these are not strong enough to ensure good solubility. Both answers A and C can undergo hydrogen bonding with water, but the alcohol will form stronger hydrogen bonds because oxygen's electronegativity is greater than nitrogen's. Therefore, the alcohol will be the most soluble.

72. (D) Compounds I and II have the highly electronegative oxygen atom in their structure that allows them to hydrogen-bond to water. Remember, hydrogen bonding involves a nitrogen atom, an oxygen atom, or a fluorine atom.

73. (C) To determine the absolute configuration about a carbon atom, note the lowest-priority group and assign priority to the other groups. Beginning with the highest-priority group, count toward the next highest group and end with the lowest group. Counting to the right (clockwise) is R, and counting to the left (counterclockwise) is S. The configuration about each carbon atom is independent of the configuration about the other carbon atom. In answers A, B, and D, the stereocenters are R, but the stereocenter is S in answer C.

74. (D) In answer D, all stereocenters have been inverted, allowing it to be a nonsuperimposable mirror image of the compound. That is not true of the others, because not all of the stereocenters have been inverted.

75. (A) Meso compounds are ones that contain two or more stereocenters. The compound is superimposable with its mirror image. A symmetry plane splits a meso compound. Compounds B, C, and D are superimposable, but compound A is not.

76. (D) To determine the absolute configuration about a carbon atom, note the lowest-priority group and assign priority to the other groups. Beginning with the highest-priority group, count toward the next highest group and end with the lowest group. Counting to the right (clockwise) is R, and counting to the left (counterclockwise) is S. The configuration about each carbon atom is independent of the configuration about the other carbon atom. Doing this shows that compound IV has an R stereocenter.

77. (B) The *d* and *l* are older symbols to represent (+) and (−), respectively.

78. (B) Sodium sulfate is a soluble ionic compound, so its solubility will be greater than those of the other compounds, even though they can all hydrogen-bond to water.

79. (D) The boiling point of a compound is highly influenced by the intermolecular forces between the compounds. The stronger the intermolecular forces, the higher the boiling point. Compound I can hydrogen-bond to other molecules, while compound II can use only weaker dipole-dipole forces.

80. (A) These two structures are diastereomers of each other, not mirror images. The left one is S and S around its stereocenters, while the right is S and R.

81. (C) These two compounds are mirror images of each other, making them enantiomers.

82. (A) N1 has only single bonds (three of them); therefore, it is sp^3 hybridization. N2 has a double bond and a single bond, which indicates sp^2 hybridization. Each nitrogen atom has a lone pair (not shown).

83. (A) There are three chiral carbon atoms ($n = 3$) on the ring containing the oxygen, so there are $2^n = 8$ stereoisomers.

84. (C) To determine the absolute configuration about a carbon atom, note the lowest-priority group (in both of these cases an H) and assign priority to the other groups. In this case, the highest-priority group on each C is the OH group, followed by the COOH group. Beginning with the highest-priority group (OH here), count toward the next highest group (COOH here) and end with the lowest group. Counting to the right (clockwise) is R, and counting to the left (counterclockwise) is S. The configuration about each carbon atom is independent of the configuration about the other carbon atom. Doing this indicates that both carbons are R.

85. (B) The side chain is a sec-butyl group, and alkyl groups are nonpolar.

86. (A) The optical rotation might be changed or might not be changed; therefore, it is impossible to determine the effect without more information.

87. (A) To determine the absolute configuration about a carbon atom, note the lowest-priority group (in the first and third carbons, it is an H, and in the second case, it is an ethyl group) and assign priority to the other groups. In this case, the highest-priority group on each C is the R group containing the ether group followed by the OCH_3 group. Beginning with the highest-priority group, count toward the next highest group (OCH_3 here) and end with the lowest group. Counting to the right (clockwise) is R, and counting to the left (counterclockwise) is S. The configuration about each carbon atom is independent of the configuration about the other carbon atom. Doing this indicates that both carbons 1 and 2 (counting from right to left) are R and the third carbon is S.

88. (D) The molecules are the same. Rotating the right-hand structure until the $-OCH_3$ is vertical yields the same structure as the left-hand structure.

89. (C) Meso refers to a plane of symmetry in an enantiomer; that is not the question that was posed. The term *achiral* refers to having superimposable mirror images, again not the question here. Absolute configuration uses the R and S notation. To determine the absolute configuration about a carbon atom, note the lowest-priority group (in this case it is a hydrogen atom) and assign priority to the other groups. In this case, the highest-priority group on each C is the R group containing the carbonyl group, next is the group containing the nitrogen, and then the hydrocarbon group. Beginning with the highest-priority group, count toward the next highest group and end with the lowest group. Counting to the right (clockwise) is R, and counting to the left (counterclockwise) is S. The configuration about each carbon atom is independent of the configuration about the other carbon atom. Doing this, the direction of rotation is to the left (counterclockwise), and so it is S.

90. (A) These two molecules are identical. Rotating the right-hand structure allows you to place it over the left-hand structure.

91. (D) There is no difference between the reactivity of the isomers (eliminating A and C). Stability in this case is related to the proximity of the groups to each other (steric hindrance). The most stable isomer will be the one in which the groups are as far away from each other as possible—the trans form.

92. (A) In order to convert from one isomer to another, the double bond (π bond) would have to be broken. This is not easy to do, as π bonds are relatively strong.

93. (C) 1,3,5-hexatriene contains three double bonds (triene), and even though there is resonance stabilization, it will undergo hydrogenation. The other compound, 1,3,5-cyclohexatriene, is benzene, which is resonance stabilized and does not hydrogenate easily.

94. (C) Meso compounds have stereocenters, but they also have symmetry, which allows them to be the mirror images of themselves. This may occur when you have at least two stereocenters.

95. (A) R and S refer to configurations, not rotations of plane-polarized light. If the light is being rotated to the left (counterclockwise), it is levorotatory; and if it is being rotated to the right (clockwise), it is dextrorotatory.

96. (C) The R form will rotate in the opposite direction from the S. The magnitude of the rotation will be the same.

97. (A) Enantiomers have the same physical properties, except for their ability to rotate plane-polarized light. Therefore, their melting points will be the same.

98. (A) The difference between a carboxylic acid and its sodium salt would be the reaction with a base to form the carboxylate ion. This does not affect the chiral center and cannot affect the rotation.

99. (D) There are four chiral centers ($n = 4$) in this molecule. The number of possible isomers is 2^n, or 2^4 in this case; $2^4 = 16$.

100. (A) The number of meso forms does not affect the number of stereoisomers, just the number of chiral centers. The number of isomers may be calculated by the formula 2^n.

101. (B) Diastereomers may have different physical properties, but their molecular formulas must be the same.

102. (D) Enantiomers have opposite configurations and optical rotations. Therefore, compound C would have an S absolute configuration and a negative optical rotation.

103. (D) The two compounds are cis/trans isomers (or diastereomers), and as such, they are expected to have different physical properties altogether. There are differences in one chiral center, so there would be a difference in the optical activity.

104. (C) To determine the absolute configuration about a carbon atom, note the lowest-priority group (in both carbons in the ring, it is a hydrogen) and assign priority to the other groups. In the case of both carbons, it is the nitrogen group, followed by the isopropyl. Beginning with the highest-priority group, count toward the next highest group and end with the lowest group. Counting to the right (clockwise) is R, and counting to the left (counterclockwise) is S. The configuration about each carbon atom is independent of the configuration about the other carbon atom. Doing this, the direction of rotation is to the left (counterclockwise), and so it is S for both.

105. (D) In order to have hydrogen bonding, there must be a hydrogen atom attached to an extremely electronegative element: O, N, or F. The only structures that meet that criterion are the alcohol and the amine, III and IV.

106. (A) To determine the absolute configuration about a carbon atom, note the lowest-priority group (in both of these cases an H) and assign priority to the other groups. In this case, the highest-priority group on each C is the Br group, followed by the ethyl, and then the methyl group. Beginning with the highest-priority group (Br here), count toward the next highest group (ethyl here) and end with the lowest group. Counting to the right (clockwise) is R, and counting to the left (counterclockwise) is S. The configuration about each carbon atom is independent of the configuration about the other carbon atom. Doing this indicates that structure I is R and structure II is S.

107. (B) Simply examining the chlorine atoms is sufficient to eliminate most of the answers. The chlorine atoms are in different positions relative to the aldehyde group, as one is 3-chloro and the other is 4-chloro. Therefore, the two compounds are not identical, conformers, or enantiomers, which leaves regioisomers.

108. (A) Meso compounds have stereocenters, and there is a plane of symmetry splitting the molecule. In I and IV, there is no plane of symmetry present. In II, there is a plane of symmetry between the two chiral centers. In III, there is a plane of symmetry cutting the molecule in two.

109. (D) Compound I can form two hydrogen bonds, compound II cannot form any hydrogen bonds, and compound III can form one hydrogen bond. The compound with the greatest number of hydrogen bonds will have the highest boiling point, and so on.

The carbonyl group adds to the attraction between the molecules, but not enough to compensate for the second hydrogen bond in the diol.

110. (C) The boiling point is related to the strength of the intermolecular forces; the stronger the forces, the higher the boiling point. Structure I can undergo strong hydrogen bonding, whereas structure II has weaker dipole-dipole forces.

111. (C) All the compounds except II have different substituent groups on the ring system. However, II has the same groups, and this leads to a plane of symmetry (the cyclobutene ring)—the plane of the molecule.

112. (B) Achiral compounds are mirror images that are superimposable. Answer B has an internal plane of symmetry, which allows it to be superimposed on its mirror image. Answer B is a meso compound.

113. (A) The 0° optical rotation indicates an achiral compound, a meso compound, or a racemic mixture. If the substance were achiral, then when reacting with an S stereoisomer, only one compound would form. The reaction of a meso compound with an S stereoisomer would yield a racemic mixture with a single melting point. If the substance were a racemic mixture, then the reaction with an S stereoisomer would yield two diastereomers with different melting points that could be easily separated. The cis isomer would probably give only one product.

114. (D) There are eight chiral carbon atoms ($n = 8$) present, so the number of optical isomers would be $2^8 = 256$.

115. (B) The (+) and (−) are optical rotations, not absolute configurations, so answers C and D can be eliminated. To determine the absolute configuration about a carbon atom, note the lowest-priority group and assign priority to the other groups. Beginning with the highest-priority group, count toward the next highest group and end with the lowest group. Counting to the right (clockwise) is R, and counting to the left (counterclockwise) is S. The configuration about each carbon atom is independent of the configuration about the other carbon atom. In this case, it would be counting counterclockwise (S).

116. (D) Meso compounds are ones that contain two or more stereocenters that are superimposable with the compound's mirror image. Answer C has no stereocenter. Answer D has stereocenters and a plane of symmetry between them, which allows it to be superimposable with its mirror image.

Chapter 3: Substitution and Elimination Reactions

117. (B) Answers C and D would be intermediates associated with an S_N1 reaction, not an S_N2. Answers A and B are S_N2; the nucleophile is negative (or δ−), eliminating answer A.

118. (B) Structures B and C are primary alkyl halides, which favor an S_N2 mechanism because there is less steric hindrance; however, Br is a much better leaving group than F. The tertiary alkyl halide, answer A, will react by an S_N1 mechanism. The secondary alkyl halide, answer D, will partially react by an S_N2 mechanism; however, Br is a better leaving group than Cl.

119. (D) The "2" in S_N2 refers to second-order kinetics. S_N2 reactions are basically one-step reactions with a five-coordinate intermediate, resulting in a substituted product. However, there is no carbocation formation—that happens in S_N1 reactions.

120. (D) Since the solvent lowered the reaction rate, it probably was interacting with the nucleophile by solvating it, making it more difficult for it to attack an electrophile. The solvent should cause none of the other changes listed.

121. (A) In S_N2 reactions, second-order kinetics applies. The rate is first-order with respect to the concentration of the substrate and first-order with respect to the concentration of the nucleophile.

122. (D) S_N2 reactions obey second-order kinetics, so the concentration of the nucleophile is important. There is inversion, so there is no retention of configuration, and S_N2 reactions are a one-step process, not two. However, primary alkyl halides do react faster than either secondary or tertiary alkyl halides.

123. (A) S_N2 mechanisms are very susceptible to steric interference. Smaller is better. The remaining factors should be the same for the two primary alkyl halides.

124. (A) Larger groups and more branching will lead to steric problems, which is particularly noticeable in an S_N2 mechanism.

125. (D) There is only one hydrogen atom to be replaced for substitution to yield the 2-chloro-2-methylpropane and nine hydrogen atoms that can be replaced to give the 1-chloro-2-methylpropane. Therefore, (36%/1 H):(64%/9 H) ≈ 5:1.

126. (A) One Li leaves with the halogen, and the other replaces the halogen on the ring.

127. (B) The alkoxide has a lone pair of electrons. Therefore, it acts as the nucleophile and attacks the most positive carbon atom. The most positive carbon atom is the one with the alcohol –OH attached.

128. (C) There are four different, nonequivalent carbons on 2,4-dimethylpentane in which monochlorination can take place.

129. (A) The starting structure is a primary alcohol, and the reactants will cause oxidation of this alcohol to yield a carboxylic acid. The conditions are such that partial oxidation to the aldehyde will not the yield the final product, as the aldehyde will also oxidize under these conditions.

130. (B) The reactant is a secondary alcohol, and the oxidation of a secondary alcohol yields a ketone. The other products will not form under these conditions.

131. (A) The organic reactant is a tertiary alkyl halide. Tertiary alkyl halides react by an S_N1 mechanism, so the bromide leaves, with the formation of the carbonium ion shown in answer A. Answers B and C are S_N2 mechanism intermediates, which are not important for tertiary alkyl halides. Answer D is a less stable secondary carbonium ion, and the tertiary carbonium ion will not rearrange to give a less stable carbocation.

132. (C) Ammonia is the best leaving group because it is a stable neutral molecule. The others would have an attraction between the anion and the group left behind.

133. (D) The leaving group in answer D would be N_2. Diatomic nitrogen, N_2, is the most stable, and it is also neutral, unlike the other leaving groups.

134. (A) In the propagation step, you have the formation of free radicals beginning with a free radical. Answer D is an initiation step (creating a free radical from a molecule), and answers B and C are both termination steps (combining two free radicals to yield a molecule).

135. (C) Tertiary free radicals are more stable than secondary, primary, or aromatic ones. In addition, this free radical is also resonance stabilized.

136. (D) A chlorination reaction is a free radical process. The intermediates should be free radicals, which eliminates answers A and C. Answer B, without rearrangement to answer D, would not give the indicated product. The replacement of the tertiary hydrogen would proceed by the formation of a stable tertiary free radical.

137. (C) Only primary and secondary alcohols undergo oxidation; answers A and D are primary alcohols, and answer B is a secondary alcohol. Answer C is a tertiary alcohol and will not undergo oxidation by acidic $K_2Cr_2O_7$.

138. (A) Secondary and tertiary alcohols react with strong acids via S_N1 reactions. I and III are tertiary alcohols. II and IV are primary alcohols.

139. (B) After dehydration of the 2-methyl-1-pentanol, there is a rearrangement to give a more stable carbocation intermediate, which leads to II.

140. (D) Free radical bromination is VERY selective—it strongly prefers tertiary carbon atoms over secondary carbon atoms, and it strongly prefers secondary carbon atoms over primary carbon atoms. Answers A and C involve primary carbon atoms. Answer B involves a secondary carbon atom. Answer D involves a tertiary carbon atom.

141. (B) Structure II is more stable than structure I because of steric considerations. In structure II, the two groups are equatorial, minimizing the repulsion.

142. (A) In order for a substance to be a nucleophile, it should (1) have a center of high electron density. Therefore, a substance with a negative charge is a stronger nucleophile than a neutral species or one with a positive charge. (2) The lower on the periodic table the element with the charge is, the stronger a nucleophile it is. Therefore, AsH_2^- is a better nucleophile than PH_2^-.

143. (A) The stronger base will have the higher electron density on the oxygen atoms on the carbonyl carbon. (The two oxygen atoms are equivalent due to resonance.) The chlorine atoms on the trichloroacetate ion tend to withdraw electron density from the oxygen atoms (inductive effect); this leaves the oxygen atoms with a lower electron density than the acetate ion. This makes the trichloroacetate ion a weaker base than the acetate ion.

144. (A) Carbocation stability is related to the degree of alkyl substitution on the carbon on which the positive charge is located. The higher the degree of substitution, the more stable the ion is. Structure I is a primary carbocation, structure II is a tertiary carbocation, and structure III is a secondary carbocation. The least stable is structure I, followed by structure III, and the most stable is structure II.

145. (B) Carbocations exhibit sp^2 hybridization and a trigonal planar geometry in order to minimize electron repulsion.

146. (A) Inversion of configuration takes place with S_N2 reactions (eliminating answer B), and there are changes in conformation (eliminating answer C). The intermediate in S_N1 reactions with chiral compounds is an achiral structure, which may be attacked from either side to give a racemic mixture and not an achiral product.

147. (B) The "1" in E1 refers to first-order kinetics. The elimination leads to the formation of a double bond, which is the formation of a π bond. The E1 mechanism is a two-step mechanism. The first step is the formation of the transition state followed by the formation of the alkene. There is no attack by another species, certainly not a nucleophile.

148. (C) In the E2 reaction, there is deprotonation by a base, loss of the leaving group, rehybridization of the reacting carbon center from sp^3 to sp^2 hybridization, and the formation of a double bond (π bond). There is no carbocation intermediate as occurs in an S_N1 mechanism.

149. (A) Alkanes are not very reactive. Halogenations (chlorination) of an alkane take place by a free radical mechanism including initiation, propagation, and termination steps.

150. (C) The reaction involves a tertiary alkyl halide, so the mechanism is S_N1 and will remain S_N1. Optical activity is not related to the rate of reaction. The stability of the product is a thermodynamic argument, not a kinetic argument. The rate depends on the reactants and the activated complex (activation energy). The reason that the rate decreased was that the activation energy for the reaction increased because the larger R groups created more steric hindrance. Watch out for mixed kinetic-thermodynamic problems on the MCAT. In general, it is not possible to use kinetics to explain a thermodynamic question or to use thermodynamics to explain a kinetic question.

151. (A) The reaction of methyl bromide will follow an S_N2 mechanism. For this reason, first-order kinetics is incorrect—S_N1 is not S_N2. Formation of a racemic mixture is incorrect because that would require a chiral carbon, something that is impossible with methyl bromide. S_N2 reactions have an inversion of structure, which means that the conformation is not retained. S_N2 reactions, however, exhibit second-order kinetics, because there are two reacting species in the rate-determining step.

152. (B) In a termination step, there is a combination of free radicals that stops the reaction. Answers A and C are propagation steps because a free radical forms from a free radical; answer D is an initiation step because a molecule generates free radicals.

153. (D) The "2" is S_N2 refers to second-order kinetics, not first-order kinetics. There is inversion of configuration and no rearrangement, and the reactivity sequence is as described.

154. (B) The reaction between ethyl iodide and the hydroxide is an S_N2 reaction; these reactions obey second-order kinetics: first-order with the alkyl halide and first-order with the hydroxide. Doubling the concentration of the hydroxide will cause the rate to double.

155. (A) The free radical chlorination of an alkane gives many products. It is not a very selective reaction (nonregioselective), and factors such as inductive effects and statistical factors will affect the products formed and their proportions. A number of products, not just the most stable, will form.

156. (A) Initiation (generation of the free radical) is the first step in all free radical mechanisms. The propagation step(s) occur(s) after the initiation step and before the termination step. The commencement and instigation steps are made-up terms.

157. (A) This is a tertiary alkyl halide; therefore, the reaction will follow an S_N1 mechanism. An S_N1 mechanism on a chiral center normally yields a racemic mixture.

158. (A) This is an electrophilic attack on a conjugated system, which leads to 1,4-addition. The 3-chloro product is the Markovnikov product. The other product results from resonance of the intermediate allylic cation. (The intermediate allylic cation also leads to some of the 3-chloro product.)

159. (B) In order for a substance to be a nucleophile, it should (1) have a center of high electron density. Therefore, a substance with a negative charge is a stronger nucleophile than a neutral species or one with a positive charge. (2) The farther to the left on the periodic table, the stronger the nucleophile. Therefore, NH_2^- is greater than OH^-.

160. (D) The reaction conditions lead to carbene formation from the $CHCl_3$. Answer D is the product of a carbene addition to a double bond.

161. (A) An S_N2 attack on a secondary alkyl halide requires a strong nucleophile to attack the alkyl halide. Alkoxide ions, like propoxide ion, are very strong nucleophiles.

162. (C) Answer B is incorrect, since the substitution does not occur at the least sterically hindered site. There is no chance of resonance with this structure, and it is not the site with the most $\delta+$. However, inductive effects, leading to monobromination, can stabilize the intermediate. Bromine is VERY selective, and a tertiary alkyl halide is strongly preferred.

163. (B) In free radical addition reactions, the propagation step involves a free radical (such as the chlorine free radical) reacting with the substrate (such as methane) to produce another free radical (such as the methyl free radical). An initiation step generates two free radicals forming from a molecule, and a termination step generates a molecule forming from two free radicals. Carbocations are never involved in free radical reactions.

164. (C) S_N1 reactions are first-order in the substrate concentration. Therefore, doubling the concentration of the sodium ethoxide (the nucleophile) will have no effect on the reaction rate. Changing the sodium ethoxide concentration would make a difference if this were an S_N2 mechanism.

165. (A) Neopentyl bromide, along with ethyl, methyl, and propyl bromide, is a 1° bromide and will yield a 1° carbocation (rearrangement may occur). The more stable carbocations are 2° and 3°. Isopropyl bromide is a 2° bromide, which will yield a 2° carbocation, which is more stable than a 1° one. There are no 3° options in the answers.

166. (D) Tertiary alkyl halides react exclusively by S_N1 mechanisms, and answer D is the only tertiary alkyl halide present. The question is not asking about elimination mechanisms.

167. (B) Free radical stability is related to the degree of alkyl substitution on the carbon on which the single electron resides. The least stable is the methyl group itself, then primary, then secondary, and finally, tertiary is the most stable. Structure III is a tertiary, so it will be the most stable. Structures II and IV are both secondary, so they should be approximately the same, and structure I is a primary, so it will be the least stable.

168. (A) To increase the acid strength of phenol, electron density needs to be withdrawn from the O-H bond. Weakening the bond makes it easier to break, and therefore it becomes a stronger acid. Both the cyano and the trichloromethyl groups are electron-withdrawing groups because of the high electronegativities of the nitrogen and chlorine atoms. Therefore, they will make the acid stronger.

169. (C) Configuration II, the eclipsed configuration, is the least stable because of steric interactions. Configuration III, gauche, is the most stable because of intramolecular hydrogen bonding. Configuration I, anti, is the second best in terms of stability. If there were no intramolecular hydrogen bonding, configuration I would be the most stable.

170. (A) The free radical chlorination of isopentane will yield all possible products shown because the chlorine radicals can attack any position. Free radical bromination is more selective, giving primarily one product. The distribution of the products will not be equal.

171. (C) Strong acids react with primary alcohols via S_N2 mechanisms. II and IV are primary alcohols, while I and III are tertiary alcohols. Tertiary alcohols never react via an S_N2 mechanism.

172. (A) This is a dehydrohalogenation with the more stable (more substituted) alkene forming. Structure A is the most substituted. Structure B has a five-bonded carbon atom, which will not form.

173. (A) This is a free radical halogenation, and bromine is very selective and goes to the tertiary carbon.

174. (B) The product of this reaction is formed by an S_N2 reaction with inversion of configuration and no rearrangement—structure II.

175. (B) These conditions will lead to the replacement of the OH with a Br.

176. (B) A nucleophile has a lone pair of electrons; in answers A, C, and D, there is a lone pair on the nitrogen. This is not true in answer B, since all electrons are in bonds. In addition, positive charges attract nucleophiles, and the positive charge on answer B would be repelled by a positive charge.

177. **(C)** There would be a carbocation intermediate formed in the reaction, and water, because of its polar nature, would stabilize the carbocation better than the less polar DMF solvent.

178. **(C)** The reaction of a primary alcohol with $SOCl_2$ yields an alkyl halide, answer C.

179. **(D)** Answer C is a primary alcohol, answers A and B are secondary alcohols, and answer D is a tertiary alcohol. Tertiary alcohols undergo dehydration faster than secondary alcohols, with primary alcohols being the slowest. Since two of the alcohols are secondary, it is unlikely that either will be the answer.

Chapter 4: Electrophilic Addition Reactions

180. **(A)** A Markovnikov hydration begins with the formation of the most stable carbonium ion. A tertiary carbonium ion will be more stable than a secondary or a primary one. The formation of a carbanion will not occur.

181. **(D)** The addition reaction of a strong acid to an alkene in the presence of peroxides is a two-step process with the formation of a carbocation intermediate. The first step is the initiation step with the formation of a free radical. Because of this free radical mechanism, the product is anti-Markovnikov.

182. **(A)** This combination of reactants is a typical Markovnikov situation. The chlorine adds to the double bond, and the hydrogen can add either syn or anti to the carbon atom, which already has the greater number of hydrogen atoms.

183. **(D)** This reaction will involve the formation of a carbonium ion, and a tertiary carbonium ion will be the most stable.

184. **(C)** These two compounds, p-toluenesulfonic acid and o-toluenesulfonic acid, are constitutional (structural) isomers, since they differ in where a group is attached. Anomers refers to the Haworth projections of carbohydrates. Confomers refers to different views of the same compound. Diasteriomers refers to optically active compounds.

185. **(A)** During this reaction, you go from three double bonds to one, which is a loss of two π bonds. The two reactants are now connected by two new σ bonds.

186. **(D)** The intermediate will be a carbonium ion, since you are starting with a conjugated diene. This will be a 1,4-electrophilic addition reaction.

187. **(B)** HI is the strongest acid of those shown. The strongest acid gives the fastest reaction because protonation of the alkene is the rate-determining step.

188. **(A)** This is an elimination reaction (E2) with a hindered base (t-butoxide). There will be no substitution product C or D. Products A and B are both E2 products. The hindered base leads to the preference for product A over product B (Hofmann's rule over Zaitsev's rule).

189. (D) Structure I has an activating group attached, and structure III has a deactivating group attached. The presence of an activating group means that compound I should react the fastest. The presence of a deactivating group means that compound III should react the slowest. Compound II, with neither an activating nor a deactivating group, should be intermediate.

190. (B) The ketone adjacent to the ring is a meta-director, so the product will be answer B. (In addition, o,p-directors tend to produce two products unless there are serious steric problems.)

191. (C) This is an addition reaction that should be a Markovnikov addition, as shown in answer C. Answer A is a free radical product and can be eliminated, answer B is the result of an anti-Markovnikov addition and can be eliminated, and answer D would not be a product of this reaction.

192. (A) Under these conditions, anti addition will occur. The result of the anti addition will be a racemic mixture. The meso product would be the result of syn addition. The other two products would not result from the reaction of bromine with a double bond.

193. (D) The mCPBA forms oxacyclopropane rings, starting trans and ending trans. Answer D is the trans product expected, and answer B is the cis product.

194. (A) The OsO_4 promotes vicinal-syn-dihydroxylation of double bonds, resulting in compound A. Answer D is the anti compound.

195. (B) The first step is hydroboration, and the second step is oxidation to the alcohol. The reaction combination gives an anti-Markovnikov addition. The presence of a boron compound is a strong hint that anti-Markovnikov addition will occur. The use of borohydride, BH_4^-, as a reducing agent may be an exception.

196. (C) Ozonolysis cleaves the double bond and gives carbonyl groups. The other product would be formaldehyde.

197. (A) Hydrogenation of an alkyne using a Lindlar catalyst yields a cis alkene. A hindered catalyst inhibits complete hydrogenation to the alkane. It is also unlikely for there to be a ring closure reaction.

198. (C) The reactant combination will give the sulfonation of an aromatic ring, which eliminates answers A and B. The N-containing group (amide) is para-directing, so compound C (the para isomer) is formed.

199. (D) This is a Friedel-Crafts acylation reaction. The isopropyl group is an o,p-director, but steric hindrance is great at the ortho position, so the para product predominates over the ortho product. Any attack on an alkyl group is unlikely except as covered under the reactions of alkanes.

200. (B) Hückel's rule states that in cyclic systems, the compound is aromatic if there are $4n + 2$ π electrons. Answer A has 14 π electrons ($n = 3$) and is aromatic; answer C has 18 π electrons ($n = 4$) and is aromatic; answer D has 6 π electrons ($n = 1$) and is aromatic; answer B has 16 π electrons, which does not fit Hückel's rule, and therefore is not aromatic.

201. (A) By Hückel's rule, a cyclic conjugated polyene may be aromatic only if it contains $4n + 2$ π electrons. This is true of compounds B ($n = 1$), C ($n = 1$), and D ($n = 1$), but compound A contains only a $4n$ π system ($n = 2$) and therefore is not aromatic.

202. (B) The Br adds as Br⁺ (followed by H_2O attacking to form the alcohol), and the Markovnikov product forms. The Br⁺ behaves like H⁺ when it comes to addition reactions.

203. (C) The mixture of nitric acid and sulfuric acid will nitrate (add NO_2 to) an aromatic ring. The –COOH is a meta-director (eliminating answers A and B). Both –COOH and –NO_2 are deactivating.

204. (A) The nitro group is a meta-director, so substitution will occur in the meta position.

205. (A) HBr in the presence of peroxide will add across the double bond in 1-pentene in an anti-Markovnikov fashion to give I as the product. Structure II is the Markovnikov product.

206. (B) A Lindlar catalyst leads to partial, not complete, hydrogenation, so an alkene will form over an alkane. The other compounds should not form.

207. (A) The reaction of meta-chloroperoxybenzoic acid (mCPBA) with 4-methylcyclohexene gives a diol through anti addition; in this case, an enantiomeric pair of trans diols forms.

208. (C) Permanganate will react with the cyclohexene to form a diol through syn addition.

209. (C) These compounds are nonsuperimposable mirror images of each other. They are enantiomers.

210. (A) A racemic mixture will form. There will be equal amounts of the two enantiomers. The two enantiomers rotate plane-polarized light in opposite directions and in the same amount. Therefore, the individual rotations of the two forms cancel to give a specific rotation of 0°.

211. (D) The addition of HBr to an alkene is a two-step process with the formation of a carbocation. It is not a free radical reaction, and the addition is a Markovnikov addition.

212. (A) The aromatic system is very stable and acts as a poor nucleophile in electrophilic substitution reactions.

213. (C) A catalyst allows a reaction to proceed at faster rates than the uncatalyzed reaction. This is the case here. In general, the hydration will occur without the acid, but much more slowly. Remember, catalysts lower the activation energy (eliminating B). The heat of reaction is a thermodynamic argument, which, in general, does not apply to kinetics (eliminating D).

214. (C) The structure of TNT is a toluene ring with three nitro groups (-NO$_2$) attached. All the groups are the same, so that eliminates answers B and D. The N=O bond is a double bond (eliminating answer A, as sp^3 hybridization cannot be part of a double bond).

215. (B) In order to end up with the specified product, one must add a hydrogen atom and a bromine atom across the double bond. The bromine is ending up on the more substituted carbon. This is a Markovnikov addition. HBr will accomplish this.

216. (A) A nitro group attached to an aromatic ring tends to withdraw electron density from the ring, deactivating it. In order to add second and third nitro groups, you must increase the temperature to give the species more kinetic energy to overcome the activation energy.

217. (A) Letting the reactant be the solvent eliminates competition from other substances. All the other choices would compete with the chloride.

218. (D) This compound has the strongest electron-withdrawing group adjacent to the double bond. The presence of electron-withdrawing groups increases the strength of the dienophile.

219. (C) Loss of a chloride ion leaves a carbocation that is stabilized by inductive effects; this is then followed by meta attack.

220. (B) Syn addition will occur under these anti-Markovnikov reaction conditions.

221. (A) Reaction with the sulfuric acid leads to the generation of the carbonium ion, which will rearrange to a more stable carbocation.

222. (C) All the alkyl groups are ortho-para-directors. The larger the group, the greater the steric hindrance at the ortho position; this leads to more para product.

223. (D) Ozonolysis cleaves the double bonds, separating the top of the molecule from the bottom. Since the two halves are identical, there is only one product.

224. (D) Relative to benzene (answer A), the methoxy group (answer D) is activating (faster reaction), while the aldehyde (answer B) and the acid (answer C) are deactivating.

225. (A) Resonance induced by electron-withdrawing groups leads to the withdrawal of electron density from the ring, which, in turn, leads to the ring being electron poor and less of a target for electrophiles. Lowering the electron density leads to the system becoming a target for nucleophiles.

226. (C) The other compounds will produce the following: compound A = primary alcohol, compound B = an acid-base reaction to yield a salt, and compound D = tertiary alcohol.

227. (D) Looking at the rate law, it is apparent that the reaction is second-order: first-order with respect to the alkene and first-order with respect to the hydrogen halide. Answer C is correct about the rate-determining step. The rate law does not indicate anything about the number of steps in the mechanism.

228. (A) In aromatic substitution reactions, the aromatic ring acts as a nucleophile due to the presence of the double bonds. A dienophile has an electron-withdrawing group conjugated to the alkene. This is not the case with the aromatic ring.

229. (C) Conformers refers to rotation around a single bond, enantiomers must have a chiral center, and geometric isomers refers to orientation around a double bond or ring. The 2-nitrotoluene and 4-nitrotoluene are constitutional (structural) isomers.

230. (B) The acid-catalyzed addition of water produces an alcohol. It starts with initial protonation of the double bonds and has a carbocation intermediate, with a Markovnikov product. It is anti-Markovnikov only if diborane/peroxide is used.

231. (D) The products do form a racemic mixture, since they are enantiomers. The reaction will be second-order: first-order in HBr and first-order in 1-butene. There is no carbocation intermediate formed. However, it is not possible to separate the enantiomers by fractional distillation, since they will have the exact same boiling point. Only some stereoselective means of separation may be used.

232. (A) Ozonolysis will cleave all the double bonds, yielding four pieces. However, three molecules (formaldehyde) are the same, so only two products will form.

233. (B) The HBr addition to 1-pentene will yield the Markovnikov product, II. Product I would be the result of anti-Markovnikov addition. Only product II will form, regardless of the temperature.

234. (D) In this reaction, one Br attacks one end of the C=C bond, and then the other Br attacks. The second attack can be from either side to give both enantiomers. Therefore, a racemic mixture will result.

235. (D) This is a hydrogenation (deuterium, D_2, in this case) reaction, and these reactions produce a syn addition product.

236. (B) Hydrogenation using Na/ammonia produces a trans-alkene from an alkyne.

237. (A) This is a Friedel-Crafts alkylation to add an ethyl group. The isopropyl group is an o,p-director, but steric hindrance at the ortho position means that the para product will predominate.

238. (A) This is a Diels-Alder reaction involving a conjugated diene and a dienophile. The triple bond becomes a double bond, and the two double bonds lead to one double bond between their original positions. The final product should have two carbon-carbon double bonds.

239. (B) The reaction is S_N, the rate law is rate = k [electrophile]. Since the reaction is S_N1, changing the amount of nucleophile will have no bearing on the time the reaction takes.

240. (C) Hydration of an alkene will produce regioisomers (sometimes called positional or structural isomers).

241. (D) This is a nitration reaction. The carboxylic acid group is a deactivating meta-director, so the nitro group will add in the meta position.

242. (A) The zinc amalgam acts as a reducing agent.

Chapter 5: Nucleophilic and Cyclo Addition Reactions

243. (A) The strong base removes an H⁺ from the carbon adjacent to the carbonyl to produce an enolate ion. The enolate ion reacts with more aldehyde to form an alkoxide ion, which interacts with water to yield an aldol. Strong bases, such as hydroxide, will not protonate. The negative charge on the hydroxide ion means that it will be a nucleophile and not an electrophile. The hydroxide ion is not a free radical, so termination is not applicable.

244. (A) The strong base removes an H⁺ from the carbon adjacent to the carbonyl to produce an enolate ion. The enolate ion reacts with more aldehyde to form an alkoxide ion, which interacts with water to yield an aldol.

245. (A) The enolate ion is a strong nucleophile, and it attacks the carbonyl carbon atom to give the observed product. The presence of the negative charge means that the enolate ion is not an electrophile. The product has incorporated the enolate; therefore, the enolate cannot be a catalyst. The enolate ion has no unpaired electrons, so it cannot be a free radical.

246. (C) The carbonyl reacts with one alcohol group to give a hemiacetal, which goes on to react with the second alcohol group to yield the acetal.

247. (C) Tautomerism is the process by which an aldehyde or ketone is in equilibrium with its enol form.

248. (A) This reaction forms an imine (answer A). The other product is water (2 H from N and an O from the carbonyl).

249. (B) This is the formation of a cyclic acetal. Under acidic conditions, carbonyls and alcohols react to form acetals.

250. (C) The Grignard reagent behaves as a nucleophile when added to a carbonyl group (attacking the carbonyl carbon), giving a magnesium alkoxide, which the acid converts to the alcohol.

251. (A) This is a saponification reaction (hydrolysis of an ester under basic conditions); the other product is an alcohol (methanol).

252. (B) This reaction forms an ester through the reaction of an acid chloride and an alcohol. The alcohol is a nucleophile attacking the carbonyl carbon, followed by loss of chloride and H^+.

253. (D) Only answer D has the methyl group in the right position on the chain to yield the carboxylic acid with the methyl group in the 3-position and a five-carbon chain. The acid side is the carbonyl side of the carbonyl group, while the other side is the alcohol side.

254. (A) Only primary alcohols will oxidize in this fashion to carboxylic acids. Secondary alcohols (answer B) give ketones, and tertiary alcohols (answer C) and phenols do not react. Only answer A is a primary alcohol.

255. (B) The reduction with lithium aluminum hydride will produce an alcohol. Then reaction of the alcohol formed with an organic acid will yield the ester B. Ester C has one less carbon atom. Answer A is an ether, and answer D is an acid anhydride.

256. (B) The nitro group withdraws electron density by an inductive effect, especially in the ortho position, weakening the O-H bond and making it a stronger acid. There is also resonance stabilization of the conjugate base, which also contributes to making the acid stronger. The meta-nitro group is less effective.

257. (D) The bromide is a better leaving group, making the compound more reactive.

258. (A) A hemiacetal has the general formula of RC(OR')(OH). Structure A fits this general formula. B is an acetal, C is an ester, and D is a diether.

259. (B) The halide ions are good leaving groups, allowing them to undergo addition-elimination reactions easily.

260. (D) Hemiacetal formation through the reaction of an alcohol with an aldehyde normally follows second-order kinetics: first-order in both the alcohol and the aldehyde.

261. (B) These conditions are typical for an aldol condensation. Compound B is an aldol condensation product.

262. (A) The reaction of an enolate anion of a carbonyl compound with another carbonyl compound is the general form of a reaction called an aldol condensation.

263. (D) The presence of the highly electronegative fluorine withdraws electron density from the carbonyl group by inductive effects, weakening the O-H bond, allowing it to leave more easily, and making this a stronger acid. The closer the fluorine is, the greater the inductive effect and the stronger the acid. Therefore, IV, in which the fluorine is on the same carbon that the carbonyl is attached to, should be the strongest acid, followed by III (fluorine attached to the adjacent carbon), then I (fluorine attached two carbons away), and finally II, the weakest acid, since there are no fluorine atoms in the structure.

264. (D) An acetal results from the reaction of an alcohol with either an aldehyde or a ketone (in the presence of acid).

265. (D) The alcohol (A) is not very acidic. For the other compounds, it is necessary to focus on the hydrogen atoms attached to the carbon atoms adjacent to the carbonyl group. The presence of the carbonyl group induces an increased δ+ on the hydrogen atoms. The presence of an adjacent alcohol is not as effective, inducing an increased δ+ on the hydrogen atoms, but it will cause C to be a stronger acid than B. The presence of two adjacent carbonyl groups, with help from the methoxy groups, makes D the strongest acid. In addition, extra stability results through resonance in the anion resulting from the loss of H$^+$.

266. (A) The conditions are typical aldol condensation conditions. Aldehyde plus acid will lead to an aldol condensation to form a hydroxy aldehyde.

267. (C) It is possible to break an amide by either acid- or base-catalyzed hydrolysis. Acid-catalyzed hydrolysis yields the ammonium ion and the carboxylic acid, and base-catalyzed hydrolysis yields the amine and the carboxylate ion.

268. (D) The reaction of an acid anhydride with an alcohol is a common procedure for producing an ester.

269. (D) There must be an α-hydrogen present. A hydrogen atom on a carbon adjacent to a carbonyl (an α-hydrogen) is acidic. Answer D has no α-hydrogen atoms; therefore, it is not acidic, and it will not react with a base.

270. (C) The conversion of a carbonyl to an alcohol is a reduction. Both answers A and C contain strong reducing agents. However, sodium borohydride (answer A), while a very strong reducing agent, is not strong enough to reduce the carbonyl. Lithium aluminum hydride (answer C) is sufficiently strong to reduce the carbonyl.

271. (D) The electron-withdrawing oxygen atoms weaken the carbon-carbon bond holding the acid group to the remainder of the compound. The closer the carbonyl is to the acid, the weaker the bond and the stronger the acid. A weaker bond facilitates the loss of carbon dioxide. Answer D has the acid and carbonyl groups closest to each other.

272. (C) The Grignard reagent attacks the carbonyl group. This is a nucleophilic attack, and it will add an ethyl group to the carbonyl carbon (giving two ethyl groups). The carbonyl group becomes an alcohol.

273. (A) Methoxide is a stronger nucleophile than propoxide, since methoxide has fewer carbon atoms available to stabilize the charge. The more stable propoxide ion is a better leaving group.

274. (B) The acid destroys the acetal, converting the "ether" groups to alcohols and giving a carbonyl.

275. (A) Two functional groups are necessary, one on each end. The two functional groups must be capable of reacting with the amine in the other compound. The acid chloride groups are the only functional groups shown that will readily react with amines. Only answer A has two reactive acid chloride groups.

276. (D) The carbonyl carbon atoms were originally connected by a C=C; therefore, loop the chain around to get the two carbonyl groups together, remove the oxygen atoms, and put a C=C in their place. The product will not have a C=C. The procedure does not yield an alcohol, answer A.

277. (A) Aldehydes react with amines to form imines, not amides. The others will yield the amide shown (heating is necessary for benzoic acid).

278. (A) Normally a strong acid will first cleave the ether to give an alcohol and an organic halide, and then the alcohol will react with the strong acid to become a second molecule of the organic halide. However, in this case, the intermediate "alcohol" is a phenol, which does not react further.

279. (D) The reaction is an acid-base neutralization reaction, which produces the salt shown.

280. (B) The difference between the reactant and the product is H_2O. Dehydration (loss of water) to form an alkene is an elimination reaction.

281. (C) If there is insufficient ammonia, the reaction stops after the acid-base neutralization step to produce the ammonium salt.

282. (A) The conversion of an alcohol to a carbonyl is an oxidation. The only choice that is an oxidizing agent is answer A.

283. (D) The conversion of a carbonyl to an alcohol is a reduction.

284. (B) This is an unfavorable resonance form because of the bad charge distribution. Oxygen is the more electronegative atom; therefore, it is the least likely to stabilize a positive charge, especially when it is adjacent to a negatively charged carbon.

285. (A) The Grignard reagent plus acid workup converts the carbonyl group to an alcohol. Answer A is an alcohol, answer B is a carbonyl, answer C is a C-H, and answer D is a C=C.

286. (C) Sodium bicarbonate is weakly basic, and it will extract only "strong" acid. The compounds are I = aldehyde, II = alcohol, and III = carboxylic acid. There are too many carbon atoms for the alcohol to exhibit significant solubility in water (or aqueous sodium bicarbonate).

287. (A) Fats are esters, and either acid- or base-catalyzed hydrolysis will break the ester group. Both methods yield the alcohol (glycerol), but acid catalysis yields the carboxylic acid, while base-catalyzed hydrolysis yields the carboxylate ion. The other reactions will not produce a carboxylate ion.

288. (D) Butyl acetate is an ester. Both A and D will react with butyl alcohol to form butyl acetate. In general, the reactivities are in this order: acid chloride > acid anhydride > acid. There is no acid chloride present; therefore, the best choice is the acid anhydride.

289. (A) The acid-catalyzed reaction between an alcohol and a carboxylic acid is the Fischer esterification and results in the formation of an ester with the elimination of water. The specific reaction here would be: $C_6H_5COOH + HOCH_2CH_3 \rightarrow C_6H_5COOCH_2CH_3 + H_2O$.

290. (C) The reaction between chloroacetic acid and ammonia is an acid-base reaction. There will be a proton donated from the acid to the ammonia (Lewis base) with the formation of an ionic compound, ammonium chloroacetate.

291. (A) You started off with three double bonds (three π bonds) and ended up with one double bond (one π bond). This means that you have lost two π bonds, and those have been converted to two single bonds (two σ bonds).

292. (B) Saponification is base hydrolysis. The base hydrolysis of an ester gives a carboxylate ion (I) and an alcohol (III).

293. (A) The acidity of the compound is related to the strength of the C-H bond and the stability of the anion produced. The hydrogen in position 4 is a hydrogen atom in the α position to each of two carbonyl groups. The two carbonyl groups increase the polarity of the C-H bond, which makes it more acidic. The loss of this H leads to resonance stabilization with each of the carbonyl groups. Therefore, this hydrogen should be most easily lost (most acidic).

294. (D) Compound I has two electron-withdrawing nitro groups, which can delocalize the charge on the conjugate base. III is stronger than II because there is resonance stabilization of the conjugate base of the ortho form.

295. (C) An acetal has $R_2C(OR)_2$, like structure III, while a hemiacetal has $R_2C(OR)(OH)$, like structure I. Structure II is neither an acetal nor a hemiacetal.

296. (D) The molar mass of acetic acid, CH_3COOH, is 60 g/mole. The experimental determination of the molar mass is twice the expected value, 120 g/mole. This is because that acetic acid exists in the vapor phase as a dimer, which the two molecules held together by strong hydrogen bonds.

297. (A) The electron density is localized on the carbonyl oxygen, as shown in the following:

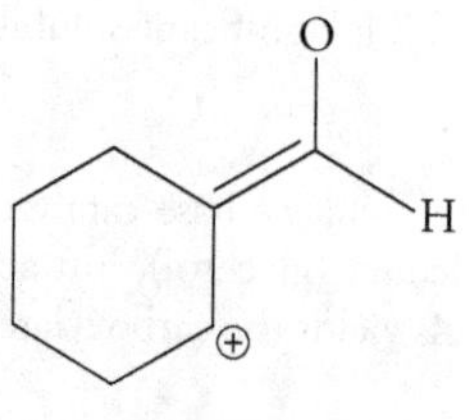

298. (D) The chloride ion is the leaving group. The water has an unpaired electron pair, making it the nucleophile, and the benzoyl chloride is seeking the electron pair, making it the electrophile.

299. (C) Treating an aldehyde with a hot base (forming a nucleophilic enolate ion) gives an aldol reaction followed by dehydration, loss of a water molecule.

300. (D) This is a Claisen condensation, in which an ester enolate attacks a carbonyl group, generating a new carbon-to-carbon bond.

301. (D) Terminal alkynes are acidic, and the Grignard is basic. The other product is propane.

302. (A) In general, the reactivities are in this order: acid chloride > acid anhydride > ester > ether.

303. (C) Pyridinium chlorochromate (PCC) is a mild oxidizing agent that will partially oxidize a primary alcohol to an aldehyde, instead of a complete oxidation to a carboxylic acid. Compound C is the only aldehyde.

304. (A) Hydrogen atoms on the carbon atom adjacent to a carbonyl group are acidic. Structure II has no carbonyl group, so it would not function as an acid at all. In structure I, there is a carbonyl group, but in structure III, the hydrogen atoms on the central carbon are adjacent to two carbonyl groups, making it the strongest acid.

Chapter 6: Lab Technique and Spectroscopy

305. (A) Because of the selectivity of bromine, the predominant product has the bromine atom attached to the tertiary carbon atom. The minor product involves attack on one of the methyl groups to give 1-bromo-2-methylpropane or $(CH_3)_2CHCH_2Br$. The methyl groups give one signal, the CH group gives another signal, and the CH_2 gives the fourth signal.

306. (A) Singlets occur when there are no hydrogen atoms on the adjacent carbon atoms to couple with the hydrogen atoms of interest. There are no hydrogen atoms on the carbon that the methyl or methylene groups are attached to; therefore, the peaks would be singlets because there is no coupling. The alcohol group gives another singlet. All other compounds have some hydrogen atoms on two adjacent carbon atoms; this leads to coupling.

307. (B) Diisopropyl ether would not have a C=C stretch. The absence of this in the IR spectrum would eliminate propene as a possible product.

308. (A) Compound I has four kinds of hydrogen that have different chemical environments (four signals), compound II has two types of hydrogen atoms (two signals), compound III also has two types of hydrogen atoms (two signals), and compound IV has only one type of hydrogen atom (one signal).

309. (D) Amines (answer A) show absorption in the 3250–3500 cm^{-1} range, alkynes (answer B) in the 2100–2260 and 3260–3600 cm^{-1} ranges, alcohols (answer C) in the 3200–3650 cm^{-1} range, and aldehydes in the 1690–1,750 cm^{-1} range (carbonyl). The correct answer is D, an aldehyde.

310. (C) This is an OH stretch. The only compound choice with an OH group is the alcohol.

311. (D) The IR indicates the presence of a carbonyl group; the NMR and Tollens' test eliminate an aldehyde, leaving only a methyl ketone to generate the iodoform.

312. (B) Sodium hydroxide would convert the boric acid to water-soluble borate ion, which ends in the aqueous layer. The organic compounds would probably be in the ether layer. Since the reaction was incomplete, some of the original ester would remain along with the alcohol formed.

313. (A) The compound is symmetric, so it is necessary to consider only one side. Starting from the left and counting to (and including) the carbonyl carbon, there are four different carbon atoms.

314. (D) There are five different chemical environments for the carbon atoms: the methyl carbons, the ring carbon between the methyl substituents, the carbons the methyl groups are attached to, the carbons that are ortho to the methyl substituents, and the carbon that is meta to the methyl groups.

315. (D) Aqueous sodium bicarbonate will extract the carboxylic acids (all the compounds). In any case, all the chemical species can hydrogen-bond to water and therefore will be soluble in the aqueous layer.

316. (A) Sodium bicarbonate will extract acids like the carboxylic acid (II); however, phenols are too weak to react. Compounds II and IV can hydrogen-bond strongly to water, and thus they would be in the aqueous layer.

317. (C) The reaction produces carbonyls (1700 cm^{-1}), not alcohols (3300 cm^{-1} and 1300 cm^{-1}) or ethers (1250 cm^{-1}).

318. (A) In general, unsaturated fatty acids have lower melting points than saturated fatty acids. The cis isomers tend to have lower melting points than the corresponding trans isomers. The lowest-melting isomer will be the one with the greatest number of cis bonds (Z): (9Z, 12Z, 15Z)-9,12,15-octadecatrienoic acid

319. (C) Extraction, chromatography, and distillation are all separation techniques, while IR spectroscopy is a technique used to determine the structure of a compound.

320. (B) NMR spectroscopy is not a separation technique; thin-layer chromatography can be used for identification, but not for an actual separation (especially if large quantities are present); and extraction would involve finding a solvent that one of the compounds is insoluble in and then removing the solvent at the end. There should be a significant difference in the boiling points of these two compounds, since the alcohol can hydrogen-bond to other alcohol molecules and the other compound cannot. Therefore, distillation should be the simplest method for separating a mixture of the two.

321. (A) For any buffer, pH = pK_a, for a mole ratio of 1.

322. (C) A racemic mixture is a mixture of enantiomers. The separation of enantiomers is called resolution. The other three answers are general laboratory techniques for separation or purification that will not separate enantiomers.

323. (B) Mass spectroscopy will confirm the chlorination, but would give little or no useful structural information in this case. IR spectroscopy and UV-Vis spectroscopy would again confirm the chlorination, but cannot easily differentiate between the possible structures. NMR spectroscopy, on the other hand, would be able to discern the differences in the structures, and, for this reason, would be the best choice.

324. (A) Filtration does not work well with many mixtures; there must be a solid involved. IR spectroscopy is not used for separating mixtures, but for structure determination. Ultracentrifugation works only if there is a very large difference in molecular weights. Chromatography can be used for separating mixtures in many cases.

325. (D) Because of the slight differences in the chemical environment of the nitro groups in the two compounds, NMR would be the best way to distinguish between the compounds. The NMR patterns for ortho and para substitution of an aromatic ring are distinctive provided the resolution is great enough.

326. (C) The toluene is nonpolar, and it does not interact with the silica gel, which is polar. Both of the nitrotoluenes, however, are polar, and they will interact with the silica gel, slowing both down with respect to toluene.

327. (A) The product would be 1,3,5-trinitrobenzene. In the 1,3,5-trinitrobenzene, there are two different types of carbon atoms: those with nitro groups attached (all identical) and those with hydrogen atoms attached (all identical). Therefore, you should observe two peaks in the ^{13}C NMR spectrum.

328. (D) The best way would be through NMR spectroscopy. The chemical environments in these two compounds are greatly different, so NMR could easily pick them out. Mass spectroscopy would not be very useful in this case, thin-layer chromatography would also not be very useful because of the lack of difference in the polarity of the molecules, and UV-Vis spectroscopy would yield a complex spectrum.

329. (B) The compound is symmetric, so it is necessary to consider only one side. Starting from the left and counting to the center (not including the carbonyl carbon, since it has no hydrogen attached), there are three different carbon atoms, giving three different types of hydrogen atoms.

330. (A) Lithium aluminum hydride converts carbonyl groups to alcohols. Therefore, the increase by four hydrogen atoms indicates that the two carbonyl groups underwent reduction. Heating an alcohol with sulfuric acid leads to dehydration. There are two alcohol groups; therefore, there are two possible dehydrations. Each dehydration reaction produces an alkene with the loss of an H_2O.

331. (D) The three methoxy groups each contain three hydrogen atoms, with no other hydrogen atoms nearby to couple with them. Compound A has only two methyl groups. In compounds B and C, at least one methyl group can couple with hydrogen atoms on the adjacent carbon atom.

332. (A) Hydrogen bonding (1-pentanol) is a stronger intermolecular force than the dipole-dipole force in 1-bromopentane. The higher molecular weight leads to a slight increase, and polymerization is not an option.

333. (A) UV-Vis spectroscopy is also known as electronic spectroscopy. Infrared spectroscopy deals with vibrating bonds. Mass spectroscopy, as the name implies, deals with the mass. NMR spectroscopy looks at nuclei.

334. (A) TLC plates are polar, and polar materials (sulfonic acids) travel less than nonpolar materials (toluene).

335. (A) Sodium bicarbonate reacts with boric acid to produce water-soluble borates. Sodium bicarbonate does not react with alcohols or esters. The alcohol and ester are more soluble in ether.

336. (D) There are 14 different types of carbon atoms (all the carbon atoms are different).

337. (B) The splitting of proton NMR signals depends upon the number of hydrogen atoms on the adjacent carbon atoms. If there are n hydrogen atoms, then the adjacent hydrogen atoms are split into $n + 1$ peaks. A triplet (three peaks) is due to two hydrogen atoms on the adjacent carbon atom.

338. (C) A band in this region is a carbonyl (eliminating answer D). Answers A and B both have two oxygen atoms; however, there is only one oxygen atom in the formula, which eliminates answers A and B.

339. (B) Deshielding shifts the signal to higher ppm values.

340. (C) The procedure will convert a carbonyl (ketone) to an alcohol (one new signal) and add a hydrogen atom to the carbon atom that is formally part of the carbonyl (a second new signal).

341. (A) Alcohols have a strong absorption in the 3200–3550 cm^{-1} region and no absorption near 1700 cm^{-1}. All the others contain a carbonyl group (1715 cm^{-1}).

342. (D) Distillation and extraction will work for large amounts of material, but the other two choices will not. Extraction with aqueous base requires an acid, but there is no acid group in either the reactant or the product. Only distillation remains.

343. (A) Water removes unreacted ammonium formate. Acid converts the amine to water-soluble ammonium ion (leaving the carbonyl compound unreacted and in the organic layer).

344. (A) Base removes hydrogen ions, leaving an overall negative charge, while acid adds hydrogen ions, leaving an overall positive charge. Positive ions move toward the negative plate, and negative ions move toward the positive plate.

345. (B) The less polar compound will migrate farther (larger R_f value) than the more polar compound. Molecular mass is much less important to TLC.

346. (D) The change in position indicates a change in the carbonyl region of the compound. HCl hydrolysis will convert an ester into a carboxylic acid with retention of a carbonyl peak. The ether has no carbonyl peak. Neither a ketone nor a carboxylic acid will change under these conditions.

347. (A) This is the iodoform test for a methyl ketone.

348. (D) Lithium aluminum hydride converts carbonyl groups to alcohols. Therefore, the increase by four hydrogen atoms indicates that the two carbonyl groups underwent reduction. No C=C underwent reduction. There is one degree of unsaturation that could be a double bond or a ring. The question asked for the maximum number of rings; therefore, there is one ring.

349. (A) Homolytic cleavage produces two free radicals. This is the initiation step in a free radical mechanism. A propagation step involves the formation of one free radical from another free radical. Instigation and proliferation steps do not exist.

350. (D) There are two types of carbons present here: six secondary and two tertiary. All of the secondary carbon atoms are equivalent, as are the tertiary carbons. Therefore, the NMR spectra should yield two peaks.

351. (C) Spin-spin coupling occurs when hydrogen atoms on one atom interact magnetically with different hydrogen atoms on an adjacent atom. In Compounds A and D, there are no hydrogen atoms on adjacent atoms. In Compound B, all the hydrogen atoms are equivalent, so there will be no coupling. In Compound C, there are hydrogen atoms on adjacent carbon atoms, so coupling should take place.

352. (B) Amines (answer D) show absorption in the 3250–3500 range; alcohols (answer C) in the 3200–3650 range; aldehydes (answer A) in the 1690–1750 range; and alkynes (answer B) in the 2100–2,260 and 3260–3600 ranges. The correct answer is B, an alkyne.

353. (A) Amines are bases; therefore, treatment with a strong acid will protonate the amine, forming a cation (ammonium) and making it water soluble.

354. (B) Isobutane has the formula $CH(CH_3)_3$. Highly selective bromination produces primarily $CBr(CH_3)_3$. The methyl hydrogen atoms are equivalent, and there are no hydrogen atoms on the adjacent carbon atom to cause splitting.

355. (D) Filtration is out; both compounds are liquids. Extraction with either acid or base is out; it is difficult to form ions with an alcohol or an ether. Fractional distillation will work, since the ether is polar with dipole-dipole forces, but the alcohol has a stronger intermolecular hydrogen bond. The difference in the intermolecular forces will make a significant difference in their boiling points, with the alcohol having the higher.

356. (A) Since there are no H atoms on the adjacent C, there will be no splitting, and a singlet will be observed.

357. (D) The formula C_nH_{2n} indicates that the compound has either one ring or one double bond. All compounds of more than three carbons with a double bond would show more than one NMR peak. Therefore, the compound must be cyclobutane, since all the hydrogen atoms would have the same chemical environment.

358. (A) In structure I, there are two types of carbons (those that are attached to an OH group and those that are not attached), so there would be two ^{13}C NMR signals. In structure II, there are four types of carbons (four signals). In structure III, there are four types of carbons (four signals). In structure IV, all the carbons are equivalent, so there would be only one signal.

359. (C) The reaction shown is the oxidation of a secondary alcohol, which yields a ketone. The absence of the O-H stretch would indicate that all the alcohol had been converted to the ketone.

360. (D) The amine group on III is basic, so extractions with dilute HCl will protonate the amine group, forming a water-soluble ion. The other materials do not react with acid, so they remain in the organic layer.

361. (A) Titration is not a separation technique, eliminating answer D. There would not be enough difference in the properties of the products to use solvent extraction. Paper chromatography is useful only for very small amounts of sample. Since there is 100 mL of mixture, fractional distillation would be the most logical method.

362. (D) IR absorption in the 1700 cm^{-1} area is indicative of a C=O bond, which all three compounds have.

363. (C) The 3600 cm^{-1} absorption in the IR spectrum is indicative of an O-H stretch. The base (OH$^-$) would be left behind in the separated water layer.

364. (A) Since there are hydrogen atoms with different chemical environments, ^1H NMR spectroscopy would be the best choice for identifying the various products.

365. (B) The aromatic system has three different hydrogen environments; therefore, there will be three resonance signals. (There are no hydrogen atoms on the carbon atoms with methyl groups attached.)

366. (A) The average of the two pK_a values is 5.4 (= pI). At the pI, the overall charge is 0; therefore, the amino acid will not migrate.

Chapter 7: Bioorganic Chemistry

367. (A) All the rings are "locked." It is necessary for the ring to open (making the aldehyde accessible) for the sugar to be reducing.

368. (D) The anomeric carbon atom in unit A is farthest to the right in the ring. The anomeric carbon atom in unit D is the "top" one (connected to unit B). An acetal is $R_2C(OR)_2$, and a hemiacetal is $R_2C(OR)OH$ (the R groups may or may not be the same).

369. (D) Each ring is one unit of unsaturation, and each double bond is another.

370. (D) Maltose is a disaccharide, but all the subunits are monosaccharides.

371. (B) Unit D is the β-anomer. The anomeric carbon in unit D is 1. The connection of unit D to unit B is to carbon 2 in unit B (numbering from the anomeric carbon 1). This combination is B.

372. (C) The structure of the hexapeptide is gly-ala-ser-gly-ser-pro. The 2,4-dinitrofluorobenzene might react with either end; however, since it is with the gly, it must react with the N-terminal amino acid.

373. (A) The gly is terminal because of the reaction with 2,4-dinitrofluorobenzene; therefore, one of the first two tripeptide fragments must be the N-terminus. There is only one ala residue; therefore, the first and last tripeptide residues must overlap, with the ala being the same in each. These tripeptides now give the sequence gly-ala-ser-gly, so the C-terminal gly must be the gly in the remaining tripeptide. Combining the third tripeptide gives the answer. The dipeptides can be used to check the prediction. All structures follow the convention of listing from N-terminus to C-terminus.

374. (A) The presence of the electron-withdrawing nitro groups makes the aromatic ring very electrophilic. The amine end of the octapeptide is a stronger nucleophile than the acid end of the octapeptide. The stronger nucleophile attacks the electrophile.

375. (B) To determine the absolute configuration about a carbon atom, note the lowest-priority group and assign priority to the other groups. Beginning with the highest-priority group (N), count toward the next highest group and end with the lowest group. Counting to the right (clockwise) is R, and counting to the left (counterclockwise) is S. The configuration about each carbon atom is independent of the configuration about the other carbon atom. Doing this, the counting is counterclockwise, or an S configuration.

376. (C) Amino acids are considered to be amphoteric because they can behave as both an acid and a base. They have a carboxylic acid end that can act as a proton donor (acid) and an amine end that can act as a proton acceptor (base).

377. (A) This is a linkage between carbon 1 (on the left-hand glucose ring) of the α-anomer and the carbon 4 on the other ring.

378. (C) A dipeptide forms when two amino acids link, with water being released. This is not the case here, since this involves the linking of monosaccharides.

379. (D) There are 6 chiral carbon atoms present in the testosterone. The number of stereoisomers can be calculated by 2^n, where n is the number of chiral carbons. Therefore, $2^6 = 64$.

380. (B) Mutarotation is the term used to define this conversion. Anomerization is an artificial term. The remaining terms refer to other changes.

381. (D) A pH of 7.4 indicates that the substance is slightly more basic than pure water, which has a pH of 7.0. A pH of 7.4 also means that the concentration of the hydrogen ion is less than the hydroxide ion concentration, so answer D is the correct answer.

382. (C) Since the leucine is immersed in a solution whose pH is the same as its isoelectric point, it will be in the zwitterion form, which has an overall 0 charge. Therefore, it will be attracted to neither electrode.

383. (A) An aldopentose is a sugar that has five carbons (pent = five Cs) and is an aldehyde (aldo = aldehyde). I and III fit those criteria; II has six carbons (hexose), and IV has no aldehyde present (ketose).

384. (D) If they formed smaller rings, there would be a great deal of strain introduced into the structure because the bond angles would be less than in a tetrahedral arrangement.

385. (C) The tertiary structure of proteins places the polar side chains on the exterior of the proteins, where they can interact with nearby molecules, such as water.

386. (B) A nucleoside is formed from a nitrogen base and a five-carbon sugar (no phosphate). A nucleoside can then react with phosphoric acid to form the phosphate ester, which is called a nucleotide.

387. (A) Acid chlorides and acid anhydrides are extremely reactive species. They are far too reactive to exist in complex biological systems.

388. (A) The structure of the micelle is a conglomerate with the carboxylate groups (which are hydrophilic, or water loving) on the exterior and the nonpolar hydrocarbon tails (which are hydrophobic, or water hating) on the interior. Because of this, the solubility of the conglomerate is increased.

389. (C) Glycine is the only amino acid that does not have a chiral carbon. This is because the α-carbon has two hydrogen atoms attached to it, $H_2N\text{-}CH_2\text{-}COOH$.

390. (B) The isoelectric point of alanine is the average of the pK_a values of the amine and carboxylic acid groups. The same two groups in ornithine will give about the same average as alanine; however, the additional $-NH_2$ significantly increases the value.

391. (A) The ammonia is a nucleophile that attacks the carbonyl carbon atom.

392. (D) In each case, the carboxylic acid group gives one pK_a. There is an additional pK_a value for each $-NH_2$ group.

393. (B) Peptide bonds (amide groups) undergo hydrolysis with acids or bases. Hot concentrated acid is better than room-temperature dilute. Acids mimic biological systems (the stomach) better than bases.

394. (A) Amino groups are typical organic bases. The carboxylic acid group is acidic, and the other functional groups are close to neutral, if not neutral.

395. (B) The pH is very basic; therefore, all acidic groups will be deprotonated. Both the carboxylic acid group and the weakly acidic phenol group will be deprotonated.

396. (C) There are four chiral carbon atoms (all but the top and bottom carbon atoms). The top half is the mirror image of the bottom half, which makes this a meso compound.

397. (A) Oxidation will convert a secondary alcohol to a ketone. A monosaccharide with a ketone group is a ketose. There are six carbon atoms (a hexose).

398. (D) Both α and β anomeric forms occur naturally, and both D and L enantiomers can occur.

399. (D) The zwitterion form has a protonated amino group (N^+) and a deprotonated carboxylic acid group (O^-). The overall charge is zero.

400. (C) In Fischer projections of carbohydrates, the key is the highest-numbered (closest to the bottom) chiral carbon. If the $-OH$ is on the left, it is L, and if it is on the right, it is D.

401. (A) The amino groups are basic, and they will react with the acid. No other group will react (be protonated). Protonation leads to a positive, not a zero, charge.

402. (B) The α-carbon (adjacent to the carbonyl) in glycine is not chiral. In all other cases, the α-carbon atom is chiral.

403. (D) Count from the oxygen atom in the ring.

404. (B) At the isoelectric point, the amino acid is in its zwitterionic form and the overall charge is zero (neutral). Decreasing the hydrogen ion concentration (increasing the pH) results in the loss of hydrogen ions (positive). Removing the positive leaves negative.

405. (A) The nonpolar (hydrophobic) side chain makes it more likely that this compound would be in the nonpolar region. All the others are polar (have S or O).

406. (A) Enantiomers will behave identically in cases B, C, and D. This leaves answer A.

407. (A) An α-amino acid has an amino group and a carboxylic acid group (not present in answers C or D) attached to the same carbon atom (the α-carbon). Answer B is a β-amino acid.

408. (C) The hydrogen atom on the chiral carbon atom will split the signal from the methyl hydrogen atoms into a doublet.

409. (D) Only the ionic forms are stable. This involves a charge at one end or the other or on both ends (the zwitterion).

410. (D) In the zwitterionic form of any amino acid, the amino group is protonated and the carboxylic acid group is deprotonated.

411. (A) The low pH will protonate the molecule, which eliminates answer D. The protonated form has a positive charge, which eliminates answer B. At the pK_a of the acid group, the equilibrium involves both the zwitterion and the protonated form (+1 charge).

412. (C) The pH is above both pK_as; therefore, both the amino group and the carboxylic acid group are deprotonated to the form shown.

413. (A) Sucrose is a disaccharide composed of a glucose and a fructose subunit. Maltose is a disaccharide composed of two glucose subunits. You need to know the common monosaccharides and disaccharides (do not worry about memorizing obscure carbohydrates).

414. (A) The compounds are nonsuperimposable, non-mirror images of each other (therefore, diastereomers), and the only difference is at the anomeric carbon, making them epimers. They are also anomers of each other, since they are epimers that differ only at the anomeric carbon. This results in two molecules that fit all the definitions.

415. (D) To determine the absolute configuration about a carbon atom, note the lowest-priority group and assign priority to the other groups. Beginning with the highest-priority group, count toward the next highest group and end with the lowest group. Counting to the right (clockwise) is R, and counting to the left (counterclockwise) is S. The configuration about each carbon atom is independent of the configuration about the other carbon atom. The carbon with the two carbons attached is not a stereocenter. For both carbons, you would count to the right (R).

416. (C) The O-P-C is typical of a phosphate ester.

417. (A) Isomerization changes one isomer to another. Isomers have the same formula. If the formula changes, it is not an isomerization.

418. (B) Hexoses (monosaccharides with six carbon atoms) can form rings in two different anomeric forms.

419. (C) Partial hydrogenation of a polyunsaturated fat requires the use of hydrogen gas with a nickel catalyst. The fat must be in excess; otherwise, complete hydrogenation will occur.

420. (B) The primary structure of a protein is the sequence of amino acids; the secondary structure involves the formation of hydrogen bonds between the protein's peptide bonds; the tertiary structure involves side-chain interactions; and the quaternary structure involves the interaction between two or more polypeptide chains.

421. (D) Hydrogen bonds are not really bonds; they are intermolecular interactions and could not hold the amino acids in a protein molecule together. Those bonds are amide bonds.

422. (B) Hydrogenation of the double bonds leads to straighter-chain molecules. Then the nonpolar chains can undergo increased London dispersion forces between the chains.

423. (A) According to the Henderson-Hasselbalch equation, pH = pK_a + log [base]/[acid]. Therefore, substituting the pH and the pK_a into the equation, you get 3.85 = 3.85 + log [base]/[acid]. The entire log term must equal zero, and the only way that this can be mathematically true is if the ratio of base to acid equals 1 (log 1 = 0).

424. (B) The pI is the average of the two pK_a values unless the side chain is also acidic or basic, which is not the case here. The average of 2.35 and 9.69 is 6.02.

425. (D) It is impossible to determine the sequence of amino acids given the results of the hydrolysis.

426. (A) This is the formation of a disulfide bond by removal of two hydrogen atoms. The oxidation of two sulfhydryl groups results in the formation of a disulfide bond. This is a redox reaction, oxidation-reduction.

427. (D) Because the straight-chain form is in equilibrium with the ring structures, as the straight-chained structure reacts, the reaction will shift to the left to produce more by Le Châtelier's principle. Therefore, all of it will eventually react.

428. (A) The resonance form would involve shifting the double bonds around the six-membered ring. Any structure that moves atoms is NOT resonance.

429. (D) Both nitrogen atoms have a lone pair (not shown). A nitrogen atom with a lone pair and three single bonds is sp^3.

Chapter 8: Final Review

430. (A) Isomers refers to different compounds having the same molecular formula. Conformers refers to rotation around a single bond. Diastereomers refers to being nonsuperimposable and not mirror images. Geometric isomers differ in the geometry around a double bond (cis/trans isomers).

431. (C) An acid anhydride consists of two carbonyl groups linked by an oxygen atom.

432. (A) This is the definition of competitive inhibition.

433. (D) The hydrogen from the alcohol adds to the center carbon of the isopropyl group (where the Mg was).

434. (A) This is the isotope effect. Heavier isotopes react more slowly than lighter isotopes. This is very important for hydrogen and less important for heavier atoms.

435. (C) Sodium bicarbonate is commonly used to extract carboxylic acids.

436. (A) This is an S_N2 situation; therefore, inversion of the configuration occurs.

437. (B) The reactants will undergo an E1/E2 mechanism, with the tertiary alcohol reacting the fastest.

438. (A) The polymer is the result of the polymerization of a substituted 1,3-butadiene. Polymerization of 1,3-butadienes leaves a C=C double bond in every repeat unit.

439. (A) Alkyllithium compounds may (rarely) act as reducing agents, but they never (in aqueous systems) act as oxidizing agents.

440. (A) Bromine reacts with trans-3-hexene to give this product.

441. (B) The bicarbonate ion, HCO_3^-, has three oxygen atoms attached to a carbon atom and the hydrogen atom attached to one of the oxygen atoms. All resonance forms MUST follow this connectivity (eliminating answers C and D). In answer A, one of the oxygen atoms and the carbon atom do not have a stable octet of electrons.

442. (A) Always begin by making sure that all forms shown are reasonable (e.g., do they all obey the octet rule?), because there are situations on the MCAT where impossible structures are present. The best (most stable) resonance form has all atoms with a zero formal charge (I). Resonance forms with the most electronegative atom (O) being negative are next in stability.

443. (A) The presence of an impurity leads to freezing-point depression.

444. (C) To be exothermic, the final state must be lower than the start (regardless of what comes between). This eliminates answers A and D. The rate-determining step has the largest E_a (greatest "jump" in energy). This is the second step in answer B and the first step in answer C.

445. (A) The activation energy to Compound I is less than the activation energy to Compound III (smaller "jump"). The lower activation energy is more rapid. The lower absolute value (Compound III) is the most stable.

446. (C) This is the "ideal" S_N2 mechanism.

447. (A) A carbon atom with three bonds and no lone pairs is sp^2. The ideal sp^2 bond angle is 120°.

448. (D) The stronger the intermolecular forces, the higher the boiling point. (These are all nearly the same size; therefore, size does not matter.) Answer A is nonpolar, which means that there are only weak London dispersion forces (lowest boiling point). All the others can form hydrogen bonds, which increases their boiling points. Hydrogen bonding involving nitrogen atoms is weaker than hydrogen bonding involving oxygen (answer C is less than answer B or D). The carbonyl group adds to the polarity of answer D, making it higher than answer B.

449. (C) Cyclohexane is nonpolar; therefore, it will preferentially dissolve nonpolar molecules (answer C) over polar molecules (answers A, B, and D).

450. (A) Alcohols are poor acids, but they are definitely better than amines (answers B and C) or alkanes (answer D).

451. (B) All the carbon atoms are equivalent. Therefore, adding one chlorine atom will always give the same product.

452. (A) Free radical bromination is VERY selective. The bromination of the tertiary carbon atom is strongly preferred.

453. (A) Cells are aqueous systems. Water reacts with acid chlorides and acid anhydrides; therefore, these species will not occur in biological systems.

454. (A) Do not forget the hydrogen atoms that are not shown. Carbon I has four single bonds (three shown plus one to a hydrogen atom), which makes it sp^3. Carbon II has two single bonds (one to a carbon atom and one to a hydrogen atom) and a double bond, which makes it sp^2.

455. (A) The key is resonance stabilization.

456. (B) The attack of the free radical on the alkene initiates the polymerization process.

457. (A) The loss of oxygen from an organic compound is a reduction.

458. (A) The loss of C=C leads to the disappearance of the 1650 cm^{-1}. The formation of the carbonyl leads to the increase of the carbonyl (near 1715 cm^{-1}).

459. (D) In order to form a coordinate to a metal, the side chain must behave as a Lewis base. To serve as a Lewis base, at least one atom needs to have a lone pair of electrons to donate. In amino acids, the atoms that can behave as Lewis bases are nitrogen, oxygen, and sulfur.

460. (A) Assigning formal charges yields a 0 or a –1 for O (overall δ–) and a 0 or +1 for the Ns (overall δ+).

461. (A) Begin by focusing on the central atom (the C). This atom has two single bonds and one double bond. This combination makes it sp^2 hybridized with a trigonal planar geometry.

462. (A) The reaction of nitric acid with sulfuric acid produces the strong electrophile NO_2^+.

463. (A) The methyl group is an ortho/para director; therefore, these isomers should result.

464. (A) Intermolecular forces affect melting points, while intramolecular forces do not. Both ortho- and meta-nitrobenzoic acid are polar molecules with no intermolecular hydrogen bonds.

465. (C) An ideal mixture should have the average of the two substances.

466. (A) In the benzene ring (initial) all carbon atoms are sp^2 hybridized. The attack changes one of the carbon atoms to four bonds and sp^3 (intermediate). The loss of a hydrogen atom from the sp^3 hybridized carbon atom gives an aromatic ring with sp^2 hybridized carbon atoms (final).

467. (A) The one with an odd number of electrons is the free radical.

468. (A) The lone electron must be in the unhybridized p orbital.

469. (C) The pH is above 2.1 and 3.9; therefore, these groups will be deprotonated. The pH is below 9.8; therefore, the amino group remains protonated.

470. (D) Amphoteric substances can behave as either an acid or a base. The carboxylic acid is the acid part, and the amine is the base part.

471. (B) Strong hydrogen bonds hold the two molecules together. All other interactions shown are weaker.

472. (A) The electron-withdrawing carboxylic acid group makes the acid slightly stronger than the alkyl groups, which are nearly identical.

473. (D) All the rings are "locked," so mutarotation cannot occur.

474. (A) In order to form a condensation polymer, two functional groups are necessary (one at each end). Answer A has only one functional group.

475. (A) The most stable structure would be the one in which both bulky side chains are in the equatorial positions.

476. (D) The side chain of asparagine is an amide (which can undergo hydrolysis); all the other side chains are hydrocarbons.

477. (A) An amine will react with an acid to form an amide. A polymer linked by amide groups is a polyamide.

Answers 263

478. (C) The free radical intermediate is planar; therefore, attack may occur from either side with equal probability. The (R) enantiomer forms from one side, and the (S) enantiomer forms from the other side.

479. (D) The molecule is symmetrical; therefore, a Cl on either end is the same compound. The unique positions (giving different compounds) are any of the methyl groups, the center carbon, and the tertiary carbon atoms (each with two methyl groups attached).

480. (A) All the carbon atoms in the ring are chiral, because each has four different groups attached.

481. (D) All double bonds contain a π bond, which gives three π bonds. All single bonds and double bonds have a σ bond. Do not overlook the hydrogen atoms, which do not appear in the figure.

482. (C) The ring gives one, and each double bond gives another.

483. (A) No carbon atoms have four different groups attached.

484. (A) Focus on the differences (sulfonic acids are VERY strong, which eliminates answer C). Phenols are weaker than carboxylic acids, which eliminates answers B and D. This leaves answer A.

485. (D) The largest groups (the methyl groups) are a maximum distance apart, and the remaining groups (H) are also not close. This is the most stable of the three staggered conformations.

486. (B) There are four different types of carbon atoms, which gives four isomers. Addition at carbon 2 or 3 (counting from the methyl carbon) can add either cis or trans relative to the methyl group to give two additional isomers.

487. (D) Free radical stability is in the following order: 3° > 2° > 1°. Answer A is primary, answers B and C are secondary, and answer D is a tertiary (the most stable)

488. (A) It is more difficult to sulfonate benzoic acid than to sulfonate toluene because the carboxylic acid group is deactivating, whereas the alkyl group is activating.

489. (C) Ammonia is a weak base, and water is amphoteric. Br is a much larger atom than Cl, so the bond to the hydrogen is weaker in HBr, making it a stronger acid.

490. (D) The balanced reaction for 1 mole of octane is $C_8H_{18} + 25/2\ O_2 \rightarrow 8\ CO_2 + 9\ H_2O$.

491. (A) The formula of methylpropane is $CH(CH_3)_3$. There are only two different sites for monochlorination to take place: on one of the methyl groups, or replacing the hydrogen on the tertiary carbon. Therefore, only two products are possible.

492. (C) The reaction between ethyl iodide and hydroxide ion is an E2 elimination to form ethane. This reaction is second-order: first-order in both ethyl iodide and hydroxide. Tripling both concentrations causes the rate to increase ninefold, $3 \times 3 = 9$.

493. (A) BH_3 has a vacant orbital that can accept an electron pair for coordinate covalent bonding. This makes it a Lewis acid.

494. (C) For polymerization, there needs to be at least two functional groups or at least one double bond. This is true of all the compounds except the 1-propanol, C.

495. (A) The methyl anion would be the strongest base, since it has a pair of unshared electrons on the carbon that can form a coordinate covalent bond with a Lewis acid. The other choices have the electron pair located on a very electronegative element, oxygen or nitrogen, which would tend to attract the electron pair, making it a weaker acid.

496. (D) Hydrogen bonding occurs when a hydrogen atom on one molecule is attached to a highly electronegative element (O, N, or F) and is attracted to an O, N, or F on another molecule.

497. (B) The left carbon atoms on each compound are achiral, and the right Cs are simply rotated. There is free rotation around the carbon-carbon single bond. The compounds are the same.

498. (A) Fluorine is more electronegative than chlorine. The fluorine atoms draw electron density away from the O-H bond by inductive effects. The reduced electron density weakens the bond, making it easier for the hydrogen to leave (stronger acid).

499. (B) The left-hand form can act only as an acid; the right-hand form can act only as a base; the zwitterion form can act as either an acid or a base.

500. (B) With HBr and peroxide, the major product will be an anti-Markovnikov addition, as is shown in answer B. Answer A is the free radical product, answer C is the Markovnikov product, and answer D looks nice.

One more. (D) In the first step, you are down to 90%; in the second step, you are at 30% of the 90% = 27%; in the third step, you end up with 70% of the 27% = 18.9%, or about 20% to one significant figure.

CPSIA information can be obtained
at www.ICGtesting.com
Printed in the USA
FSHW011720050620
70883FS